16 0129
TELEPEN
D0276722

WITHDRAWN

FISHING VESSEL SAFETY

Blueprint for a National Program

Committee on Fishing Vessel Safety
Marine Board
Commission on Engineering and Technical Systems
National Research Council

NATIONAL ACADEMY PRESS
Washington, D.C. 1991

NATIONAL ACADEMY PRESS • 2101 Constitution Avenue, N.W. • Washington, D.C. 20418

NOTICE: The project that is the subject of this report was approved by the Governing Board of the National Research Council, whose members are drawn from the councils of the National Academy of Sciences, the National Academy of Engineering, and the Institute of Medicine. The members of the panel responsible for the report were chosen for their special competencies and with regard for appropriate balance.

This report has been reviewed by a group other than the authors according to procedures approved by a Report Review Committee consisting of members of the National Academy of Sciences, the National Academy of Engineering, and the Institute of Medicine.

This report is supported by Contract Number DTCG23-89-C-200113 between the U.S. Coast Guard of the Department of Transportation and the National Academy of Sciences.

Library of Congress Cataloging-in-Publication Data

National Research Council (U.S.). Marine Board. Committee on Fishing
Vessel Safety.
Fishing vessel safety : blueprint for a national program /
Committee on Fishing Vessel Safety, Marine Board, Commission on
Engineering and Technical Systems, National Research Council.
p. cm.
Includes bibliographical references and index.
ISBN 0-309-04379-4 : $29.95
1. Fisheries—United States—Safety measures. 2. Fishing boats—
United States—Safety measures. 3. Fishing boats—Safety
regulations—United States. 4. Fisheries—Safety regulations—
United States. I. Title.
SH343.9.N37 1991
363.11′96392′0289—dc20 91-8978
CIP

Printed in the United States of America

Cover: The shrimper *John* and *Olaf* aground and breaking up, Shelikof Strait, Alaska, early 1970s. The vessel grounded on a shoal off the Alaska Peninsula and her engine room flooded while seeking refuge from a fierce storm, heavy icing and the threat of capsizing. A Coast Guard cutter attempting rescue suffered severe icing and was forced back to Kodiak. The captain and three crewmen abandoned the fishing vessel to a liferaft tethered to the rail. The raft was found ashore 70 miles away after breaking free. All four fishermen, including a father and son, were lost. Coffee cups later found on the galley table and on the bridge revealed that the vessel had weathered the storm without capsizing. (*U.S. Coast Guard*)

COMMITTEE ON FISHING VESSEL SAFETY

Committee Members

ALLEN E. SCHUMACHER, *Chairman*, American Hull Insurance Syndicate (retired)
WILLIAM G. GORDON, *Vice-Chairman*, New Jersey Marine Sciences Consortium
BRUCE H. ADEE, University of Washington
DESMOND B. CONNOLLY, Independent Marine Services, Inc.
JOHN E. deCARTERET, U.S. Coast Guard (retired)
GUNNAR P. KNAPP, University of Alaska, Anchorage
HAL R. LUCAS, Sahlman Seafoods, Inc.
JAMES O. PIERCE II, University of Southern California
LARRY D. SUND, Golden Age Fisheries
BRIAN E. TURNBAUGH, Point Judith Fishermen's Cooperative
JACK R. WILLIS, Zapata Haynie Corporation
MADELYN YERDEN-WALKER, Consultant

Government and Industry Liaisons

STEVEN C. BUTLER, Occupational Safety and Health Administration
THOR LASSEN, National Council of Fishing Vessel Safety and Insurance
NORMAN W. LEMLEY, U.S. Coast Guard
EDWARD LOUGHLIN, National Marine Fisheries Service
ROBERT C. ROUSH, NOAA Corps, National Sea Grant College Program
JOHN C. SCHERWIN, American Bureau of Shipping
DON TYRRELL, National Transportation Safety Board

Marine Board Staff

CHARLES BOOKMAN, Director
WAYNE YOUNG, Project Officer
PAUL SCHOLZ, Research Associate (through August 1990)
ANDREA JARVELA, Editor
AURORE BLECK, Administrative Assistant

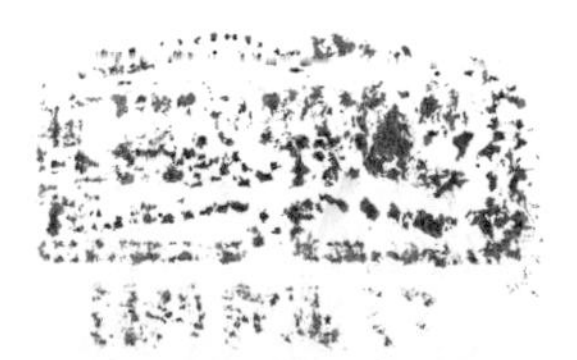

MARINE BOARD

Preface

BACKGROUND

In recent years, public awareness of fishing vessel accidents—some resulting in the loss of all on board—has focused national attention on widespread safety problems in the commercial fishing industry. Each year an average of 250 fishing vessels are lost along the Atlantic, Gulf, Pacific, and Alaskan coasts, and over 100 fishermen lose their lives pursuing their occupation. These fatalities, while perhaps low in aggregate numbers, nevertheless reflect a high rate of occurrence relative to most other occupations. Commercial fishing is widely perceived as perhaps the most dangerous occupation in the United States. Despite persistent danger, fishermen continue to fish. The foremost concern of this study is how to underpin commercial fishing with practical options for improving safety.

Congress responded to strong expressions of concern over unabated losses of fishermen and their vessels and to fishermen's concerns over rising costs of insurance with the Commercial Fishing Industry Vessel Safety Act of 1988 (CFIVSA, P.L. 100-424), included as Appendix J. Among other provisions, the act instructed the Secretary of Transportation to conduct a "Fishing Industry Vessel Inspection Study," using the facilities of the National Academy of Engineering.

In response to the congressional mandate, the secretary requested that the National Research Council (NRC) of the National Academies of Sciences and Engineering conduct a comprehensive assessment of vessel and personnel safety problems and develop a full range of safety management alternatives, including vessel inspection. In requesting the study, the secretary advised that the Coast Guard intended to use the technical information, analysis, and recommendations in a report to Congress and in its decision making concerning specific programs and regulations to improve fishing vessel safety.

COMMITTEE MEMBERSHIP

The NRC convened the Committee on Fishing Vessel Safety under the auspices of the Marine Board of the Commission on Engineering and Technical Systems. Committee members were selected to ensure a wide range of expertise and a broad spectrum of viewpoints. The principle guiding the constitution of the committee and its work, consistent with the policy of the NRC, was not to exclude members with potential biases that might accompany expertise vital to the study, but to seek balance and fair treatment. Members of the committee were selected for their experience in fishing vessel design, construction, and conversion; fishing vessel operations on the Atlantic, Gulf, and Pacific coasts and in Alaska; fishing technology development; nautical education; maritime emergency response; safety data acquisition and analysis; maritime safety analysis and enforcement; and safety management and training. One committee member is a full-time commercial fisherman. A balance of regional, academic, industrial, and government perspectives was another major consideration. Biographies of committee members are provided in Appendix A.

The committee was assisted by the American Bureau of Shipping, National Council of Fishing Vessel Safety and Insurance, National Marine Fisheries Service, National Sea Grant College Program, National Transportation Safety Board, Occupational Safety and Health Administration, and United States Coast Guard, all of whom designated liaison representatives.

SCOPE OF STUDY

The committee was asked to conduct a broad-based study of safety problems in the U.S. commercial fishing industry, including identification and characterization of safety problems. The committee was also asked to recommend general strategies for addressing the different classes of problems and, in accordance with congressional direction, to make a specific recommendation on the role of vessel inspection in improving safety. Development and preliminary analysis of individual improvement alternatives for each class of safety problem are within the scope of study. However, detailed assessment of implementation alternatives exceeds the scope of this report.

Throughout the study, the committee's principal focus was on the men and women who make up the harvest sector of the fishing industry and the vessels that are their workplace. The committee considered a broad range of safety options, some of which—if fully implemented—could potentially change the basic character of the fishing industry. The operational and structural safety of fish tender and processing vessels was also addressed. Industrial aspects of fish processing are beyond the scope of this study.

STUDY METHOD

To size the problem, the committee prepared its own description of the current commercial fishing industry—including people, boats, gear, fishing grounds, fish stocks, economic value, landings, and safety issues. This and other background papers prepared for the study are listed in Appendix B. Coast Guard casualty data were analyzed to characterize the nature, scope, and causes of safety problems. This information was compared with the agency's search and rescue (SAR) statistics. Fishing vessel compliance with safety regulations was estimated from law enforcement data compiled by the Coast Guard at the committee's request and from available Coast Guard reports. The committee also considered prior and parallel Coast Guard and industry safety initiatives and acquired and analyzed data, literature, and other reference materials from all over the world.

The committee recognized that safety problems occur on the water and are often regional in nature, and that the full character of problems in and solutions for the fishing industry could not be fully appreciated or ascertained without reaching out directly to the industry, the fishermen themselves, and entire fishing communities. Public discussion was stimulated to obtain grass-roots perspectives on problems, their causes, possible solutions, and receptivity to safety improvement strategies.

Regional assessments of fishing vessel safety were commissioned for Alaska, the West Coast, Hawaii and the Southwest Pacific, the Caribbean and Gulf Coast, the South Atlantic, and New England/mid-Atlantic. The purpose of these efforts was to develop essential information from a cross section of regional and local sources. They included:

- data on characteristics and economies of regional fisheries;
- numbers and characteristics of the commercial fishing fleets;
- numbers and types of fishermen, including qualifications and training;
- cultural information;
- descriptions of regional safety improvement resources and programs; and
- regional perspectives on safety problems and improvement alternatives and how they might be addressed by the fishing industry, government, or other organizations.

The regional assessments were publicly announced and individual viewpoints solicited through trade publications. Committee members and staff attended and led open discussions at conferences, trade shows, fishery management council meetings, trade association meetings, and similar events in order to meet personally with and obtain views about safety problems and solutions from segments of the fishing industry—including fishermen, trade associations, fishermen's wives' associations, marine surveyors, underwriters, vocational trainers, and regulators.

The committee developed an interview guide to obtain consistency in the regional assessments in order to gain insight on individual experiences and attitudes about fishing vessel safety. Information on education and training was developed through personal contact with training organizations and supplemented by correspondence with over 35 organizations having Coast Guard-approved maritime training courses.

In analyzing problems and identifying solutions, the committee drew on the expertise and insights of many experts. Their backgrounds included commercial fishing, fisheries management, insurance, personnel recruitment, safety and vocational training, anthropology and sociology, marine biology, maritime law, and epidemiology.

REPORT ORGANIZATION

Chapter 1 discusses indications of widespread safety problems, perceived safety inadequacies, and the history of safety efforts. It also identifies the objectives and major thrusts of this report, describes the assessment and solution-identification approach, and introduces the concept in which individual safety alternatives are viewed as interacting elements of an integrated safety structure.

Chapter 2 describes the commercial fishing industry—the context in which safety is considered. It examines the participants, fisheries, fisheries management system, and legislation affecting commercial fishing. Assessed numbers of fishermen and fishing vessels are tabulated regionally to establish a baseline for measuring the scale of safety problems and solutions.

Chapter 3 reports and evaluates available safety data, identifies what is known about safety problems affecting the fishing industry, and identifies alternatives for improving data.

Chapters 4 through 7 assess commercial fishing safety factors and identify and characterize possible solutions as they relate to fishing vessels and their operating equipment, survival situations and equipment, the role of fishermen in causing marine casualties, and external factors such as fisheries management practices, insurance, and weather services.

Chapter 8 blends the individual safety improvement alternatives into a comprehensive strategy and presents the committee's conclusions and recommendations.

Acknowledgments

The committee gratefully acknowledges the generous contributions of time and information provided by the liaison representatives, their agencies and organizations, and the many individuals in government, the fishing industry, marine education, and other organizations interested in improving safety aboard fishing industry vessels. Special thanks are extended to all those who communicated with the project by telephone, mail, and personal interviews.

Robert Roush of the National Oceanic and Atmospheric Administration researched and provided numerous Sea Grant reports relevant to safety in the fishing industry. Barbara O'Bannon of the National Marine Fisheries Service assembled and provided commercial fishery statistics. John Scherwin of the American Bureau of Shipping provided extensive technical support for development of Chapter 4. Dr. Bonnie McCay, Cook College, Rutgers University, presented her research on safety perspectives of New Jersey fishermen. Dr. Samuel Milham, State of Washington Department of Health, presented his use of proportional mortality rates for Washington State residents and their application to fishermen. Peggy Barry kept the committee posted on activities of the Commercial Fishing Industry Vessel Advisory Committee. Wiley Stewart, Cameron Wallace, and Christopher Hayes of the Canadian Coast Guard's Ship Safety Office provided details of fishing vessel safety activity in Canada. A. J. M. Legge and D. N. Gillstrom of the New Zealand Maritime Transport Division provided material on fishing vessel safety in New Zealand.

The support provided by many U.S. Coast Guard representatives nationwide was invaluable. Norman Lemley helped the committee clarify the scope of the project. Michael Karr provided technical assistance in developing Appendix G, the overview of vessel inspection. Thomas Purtell, Sue Holden, and William Dyson provided invaluable support in generating casualty, search and rescue, and law enforcement data, respectively. Glenn Sicks and Michael Conway provided valuable insight and data on commercial fishing and Coast Guard law enforcement in Alaska. Robert Markle and the staff of the Survival Systems Branch, Coast Guard Headquarters, provided valuable technical support for development of Chapter 6 and Appendix I. Bruce Piccard provided information on international licensing programs and the Coast Guard's development of a licensing plan.

Very special thanks are extended to the regional coordinators, through whom the committee was able to involve a broad cross section of the fishing industry in the study and develop data: North Atlantic, Dennis Nixon; South Atlantic, Robert Jones and John Maiolo; Gulf/Caribbean, Robert Jones and Dewayne Hollin; West Coast, Robert Jacobson, Ginny Goblirsch, and Fred Van Noy; Alaska, Nancy Munro and the staff of Saltwater Productions; and Hawaii/Southwest Pacific, Robert Bourke. The support of the Pacific States Marine Fisheries Commission in cosponsoring the West Coast regional assessment is gratefully acknowledged.

The extraordinary cooperation and interest in the committee's work of so many knowledgeable individuals were both gratifying and essential.

Contents

The National Academy of Sciences is a private, nonprofit, self-perpetuating society of distinguished scholars engaged in scientific and engineering research, dedicated to the furtherance of science and technology and to their use for the general welfare. Upon the authority of the charter granted to it by the Congress in 1863, the Academy has a mandate that requires it to advise the federal government on scientific and technical matters. Dr. Frank Press is president of the National Academy of Sciences.

The National Academy of Engineering was established in 1964, under the charter of the National Academy of Sciences, as a parallel organization of outstanding engineers. It is autonomous in its administration and in the selection of its members, sharing with the National Academy of Sciences the responsibility for advising the federal government. The National Academy of Engineering also sponsors engineering programs aimed at meeting national needs, encourages education and research, and recognizes the superior achievements of engineers. Dr. Robert M. White is president of the National Academy of Engineering.

The Institute of Medicine was established in 1970 by the National Academy of Sciences to secure the services of eminent members of appropriate professions in the examination of policy matters pertaining to the health of the public. The Institute acts under the responsibility given to the National Academy of Sciences by its congressional charter to be an adviser to the federal government and, upon its own initiative, to identify issues of medical care, research, and education. Dr. Samuel O. Thier is president of the Institute of Medicine.

The National Research Council was organized by the National Academy of Sciences in 1916 to associate the broad community of science and technology with the Academy's purposes of furthering knowledge and advising the federal government. Functioning in accordance with general policies determined by the Academy, the Council has become the principal operating agency of both the National Academy of Sciences and the National Academy of Engineering in providing services to the government, the public, and the scientific and engineering communities. The Council is administered jointly by both Academies and the Institute of Medicine. Dr. Frank Press and Dr. Robert M. White are chairman and vice-chairman, respectively, of the National Research Council.

Executive Summary

In every coastal region of the United States, commercial fishing industry vessels break down, are wrecked, or are lost, and fishermen are injured or die. The industry has a fatality rate comparable to those of miners, loggers, log truck drivers, and members of other high-risk occupations. Commercial fishing exposes fishermen to danger the entire time they are aboard their vessels—during transit, when fishing, and while resting. The annual toll in lives and property—on the average, over 100 deaths and 250 vessels lost—is a heavy price to pay for this small but economically important industry. The cost in terms of injuries is extensive, but poorly documented. In Alaska alone (one state that provides injury compensation for fishermen), about 1 in 20 fishermen with commercial fishing licenses requested compensation for injuries in fiscal year 1987 (FY 87). But, even this represents only a portion of the work-related injuries that occurred in the Alaska fishing industry.

Finding the fishing industry's safety record unacceptable, Congress passed the Commercial Fishing Industry Vessel Safety Act of 1988 (CFIVSA, P.L. 100-424), mandating new safety requirements across the entire fishing fleet. The act instructed the Secretary of Transportation to conduct a "Fishing Industry Vessel Inspection Study" under the auspices of the National Academy of Engineering. The act requires an assessment of safety problems and a specific recommendation on whether a vessel inspection program should be implemented.

WHAT ARE THE SAFETY PROBLEMS?

The assessment found that no single causal factor dominates fishing industry casualties. Instead, there is a complex interaction involving vessels, equipment, fishermen, the environment, and external factors such as fisheries management practices. These interactions vary, creating different situations involving life- or vessel-threatening events throughout the fishing fleet and on all fishing grounds. However, causal relationships are not well documented or understood.

The number of hull and equipment failures and related fatalities is high and varies significantly by region and vessel size. For example, material failure incidents are very high along the West Coast and North Atlantic, groundings occur relatively more frequently in Alaska, and collisions stand out as a safety problem on the Gulf Coast. The largest aggregate number of vessel casualties and fatalities involves vessels under 79 feet long. However, casualties and fatalities involving vessels 79 feet or greater in length occur at a substantially higher rate and tend to be more serious in terms of dollar losses and human costs. Estimated fatality rates increase dramatically with vessel size and, for vessels 50 feet or greater in length, exceed those of other high-risk industries. Estimated fatality rates for vessels 65 feet or greater in length are extraordinarily high, with fatalities more likely to occur from occupational causes, such as being caught in moving machinery, than from casualties to the vessel. There may also be variations based on fishery and fishing gear, although this cannot be ascertained from available data.

Human factors are implicated as a direct or secondary cause in many incidents—especially those involving capsizings, collisions, and groundings, and accidents resulting in fatalities. When vessels are exposed to sudden, catastrophic loss, whatever protective clothing fishermen are wearing is frequently their only hope for survival. Yet, most fishermen do not routinely wear protective clothing or safety equipment with inherent flotation. Even when there is enough time to orderly abandon ship, they often do not have adequate survival equipment or have not prepared themselves to don or deploy it under duress.

Other leading fishing nations—such as New Zealand, Japan, Norway, and Great Britain—use formal measures to improve professional competence among fishermen and the material condition of their vessels. In contrast, the U.S. fishing industry and government have pursued voluntary, piecemeal safety measures that lack cohesive leadership or coordination and are constrained by limited resources. While improvements to safety have been experienced on a vessel-by-vessel, person-by-person basis industrywide, voluntary measures have not achieved measurable results.

CAN SAFETY BE IMPROVED?

The committee concluded that the commercial fishing industry can be made safer by mandating systematic, industrywide attention to:

- professional qualifications;
- suitability and physical condition of vessels and equipment; and
- safe operational and occupational practices.

The committee recommends a comprehensive, integrated strategy to ensure that each safety problem is fully considered. This strategy includes developing and implementing an appropriate range of alternatives that maintain balance among other program elements and having resources available to implement them. It embodies the concept of starting with least-cost, least-intrusive mandatory measures that bring existing endeavors together into a unified program and advancing to more-stringent intervention if safety goals and objectives are not met. Safety records in different segments of the industry will need to be accommodated and will require effective monitoring to evaluate whether safety measures have the desired effect.

Specifically, the committee recommends:

- basic safety and survival training for fishermen;
- skills development for vessel operators;
- some form of certificate or license to validate that essential skills have been acquired and to motivate attention to safety; and
- an inspection program for vessels (beginning with compulsory self-inspection with audits) to ensure that they are fit for service.

The Coast Guard should lead and coordinate the program with support from other federal agencies, the fishing industry, fishery management organizations, naval architects, marine surveyors, marine educators, insurance underwriters, and others. To increase attention to safety as an element of fisheries management, the Secretary of Transportation and Undersecretary of Commerce for Oceans and Atmosphere (National Oceanic and Atmospheric Administration) should petition Congress to establish a Coast Guard flag officer as a voting member on each regional fishery management council. Congress should provide additional enabling authority and the resources needed to implement these measures. Ultimately, the level of federal and industry resources that can be committed to improving safety will be a principal determinant of the configuration of the resulting programs.

FISHING VESSEL SAFETY

1
The Need for New Approaches to Fishing Vessel Safety

Fishermen throughout the world are exposed to risks in nature like those faced by all seafarers—heavy seas, high winds, and poor visibility. In addition, they face hazards unique to their occupation—operating near dangerous shoal waters, handling moving gear on rolling decks, and opening the holds of their vessels at sea to stow the fish they catch (Pizzo and Jaeger, 1974; Murray, 1962; Yoder, 1990). Some fishermen are capable mariners whose abilities are well matched to the operating limitations of their vessels; others are not. Still others, capable enough under ordinary conditions, may be overwhelmed by circumstances beyond their control. Depressed economic conditions and fishery management regimes increase the pressures on fishermen to earn their living. They may risk fishing in foul weather during short seasons, overload boats, install gear or operate on fishing grounds for which a vessel is not designed, or simply disregard principles of good seamanship.

Though people have fished for their livelihood since earliest times (some fishing boats, gear, and methods can be traced back centuries), fishermen today must cope with changing technology and the risks that accompany it (deCarteret et al., 1980). Some innovative fishermen modify boats, gear, and deck layouts to improve their ability to harvest, especially when fishing opportunities are perceived as only fair to poor (Dewees and Hawkes, 1988; Levine and McCay, 1987; Browning, 1980). Such modifications, however, can change a vessel's operating characteristics and stability, causing great variability among even similarly designed boats and creating new safety problems.

Added to wind, waves, vessels, and equipment is the human factor. Who are these men and women who go to sea to fish for a living? How do they cope

SOME TERMS USED IN THIS REPORT

Captain is the title of the person or officer in charge of a vessel and responsible for its navigation and direction of its operation regardless of official rank or license held. Often used as a courtesy title, particularly for unlicensed individuals. In the fishing industry, denotes the person in charge of underway procedures, fishing operations, and supervision of the crew.

Certification refers to a process through which a document is issued testifying that one has fulfilled certain requirements (as of a course or school), but does not by itself impart legal permission to an individual to engage in a business, occupation, or activity.

Crew/crewmen are the body of seamen other than the captain (and licensed officers) who man the vessel. Includes processing-line workers aboard catcher/processors and processors.

Education refers to instruction to develop and cultivate knowledge. Relevant to the fishing industry, education applies generally to theory and concepts that form the foundation and framework for technical training. Frequently used together with "training" to denote the interrelationship of knowledge and practical skills.

Fishermen are the captain, licensed officers, and all members of the crew engaged in service on deck or in engineering departments aboard a fishing industry vessel. Principal occupational activities include vessel, fishing, harvesting, and delivery operations.

Individual in charge/vessel operator are generic terms used in lieu of "captain" or "master" for the person on board who is in charge of a vessel and responsible for its navigation and direction of its operation regardless of official rank or license held.

License refers to a document that imparts legal permission for a vessel to engage in a trade or activity or for an individual to engage in a business, occupation, or activity.

Licensing refers to a process leading to issuance of a license to an individual. A Coast Guard-issued license attests that certain prerequisites have been met (e.g., accumulated experience, certification of training, demonstration of skills) and that a formal examination, normally written, has been passed. The Coast Guard license, once issued, is used to fix responsibility and enforce discipline among license holders.

Master is the legal title for the merchant marine officer who is licensed and qualified for command and is serving as captain of a merchant ship. Sometimes used generically to refer to the individual in charge of a vessel.

Processing-line workers are individuals performing industrial functions aboard fish processing vessels.

Training refers to specific instruction to impart technical knowledge and develop practical skills in the application of knowledge.

Watchkeeper refers to a member of a vessel's crew responsible for operating the vessel during a period of time. On small fishing vessels, generally the person assigned by the individual in charge to operate the vessel for a specified period of time.

with the long hours and bruising working conditions? How are they educated, trained, and motivated? What do they perceive as risks? Many of these factors are poorly understood. What is known is that risks associated with human behavior are a principal cause of vessel losses and damage, and human fatalities and injuries. More specific discussion of the dangers of commercial fishing, their causes, and what may be done to overcome them follows in ensuing chapters. First, we look at the magnitude of safety problems in the commercial fishing industry and what has been done to try to alleviate them.

COMMERCIAL FISHING ACCIDENTS

Vessels and Lives Lost

Evidence of safety problems exists throughout the commercial fishing industry—inshore and offshore—and on board about 30,000 fishing vessels "documented" by the U.S. Coast Guard and approximately 80,000 registered and "numbered" by the states (see box). The Coast Guard annually investigates about 1,100 fishing vessel casualties. These are incidents involving vessel damage exceeding $25,000, a fatality, or serious injury. The agency's main casualty (CASMAIN) records for federally documented fishing vessels indicate that between 1982 and 1987, 6,558 reported casualties resulted in 1,298 vessel total losses and nearly $378 million in damages. At the same time, the Coast Guard recorded 648 commercial fishing fatalities; 439 were vessel related (associated with capsizing, fires, groundings, or collisions) and 209 were non-vessel related (falls on deck, man overboard, or people entangled in machinery).

During fiscal years 1982-1987 (FY 82-87), the Coast Guard also recorded on average over 3,100 search and rescue (SAR) cases each year for commercial fishing vessels. Over 80 percent of them occurred within 20 nautical miles of the coast and 50 percent within 3 nautical miles of shore or in inland waters. In FY 88, over 4,400 fishing vessel SAR cases were recorded. The Coast Guard expends significant federal funds annually responding to SAR cases involving

FISHING VESSEL CLASSIFICATIONS

Fishing vessel is a vessel that commercially engages in the catching, taking, or harvesting of fish or an activity that can reasonably be expected to result in the catching, taking, or harvesting of fish. (46 U.S.C.A. §2101(11a))

Fish processing vessel under federal regulations means a vessel that commercially prepares fish or fish products other than by gutting, decapitating, gilling, skinning, shucking, icing, freezing, or brine chilling. (46 U.S.C.A. §2101(11b)) At the state level, fish processors are treated more broadly and include preparations that freeze products without boxing them.

Fish tender vessel means a vessel that commercially supplies, stores, refrigerates, or transports fish, fish products, or materials directly related to fishing or the preparation of fish to or from a fishing, fish processing, or fish tender vessel or a fish processing facility. (46 U.S.C.A. §2101(11c))

Fishing industry vessel refers to any or all of the above—fishing, fish processing, or fish tender vessels, including those that have combined or convert between uses.

Inspected vessels are those subject to statutory certification under Title 46 of the U.S. Code, in which certain vessels must be certificated by the Coast Guard before they may be legally operated. A formal inspection for compliance with design, construction, and operating equipment standards is required. It applies to fish processing vessels over 5,000 gross tons and fish tender vessels over 500 gross tons. A Certificate of Inspection (COI) is issued upon satisfactory completion of the inspection. *The COI is accepted as compliance with applicable laws and regulations*, barring significant deficiencies.

Uninspected vessels are commercial vessels not subject to formal Coast Guard inspection. Uninspected fishing industry vessels make up over 99 percent of the national fishing fleet. Uninspected fishing industry vessels over 200 gross tons are required to have licensed officers. One purpose of this study is to provide technical information to the Coast Guard to determine whether formal inspection should be required for all or parts of the uninspected fishing industry fleet.

Documented commercial fishing industry vessels are those with admeasurement of at least 5 net tons for which a Certificate of Documentation has been issued by the Coast Guard.

> **State-numbered** commercial fishing vessels are those undocumented vessels under 5 net tons (approximately 32 feet or smaller), which are registered with the states under a federally prescribed numbering system. In this report, state numbered also refers to vessels bearing Alaska numbers that are administered by the Coast Guard, since Alaska does not have a state numbering system.
>
> SOURCES: 46 U.S. Code Annotated; U.S. Coast Guard Marine Safety Manual Volume III.

commercial fishing vessels. Incidents in which emergency assistance does not involve the Coast Guard are not recorded, but the number is believed to be significant.

Personal Injuries

Nonfatal injuries are a serious safety issue because of the human suffering injuries cause and the financial burden they represent to fishermen, the industry, and society (Nixon and Fairfield, 1986; Rice, MacKenzie, and Associates, 1989). The full nature and scope of injuries are often recorded only partially or not at all. Available statistical data only hint at the problem. Coast Guard injury records—though incomplete—totalled 13,916 injuries nationwide between 1982 and 1987, a yearly average of 2,319. Alaska alone (one of the few states that provide injury compensation for fishermen) recorded 2,363 personal injury claims in FY 87, representing about 1 in 20 fishermen eligible for assistance under the state's Fishermen's Fund. Compensation for these claims has an upper limit of $2,500 per individual per claim (Alaska Department of Labor, 1988).

Defining the Population at Risk and Determining Who Is Responsible

Primary responsibility for safety traditionally rests with those at risk. Government generally intervenes only when those at risk are unable or unwilling to effectively address safety in their work environments. Federal laws, regulations, and court decisions definitively allocate responsibility for safety of uninspected fishing vessels. Owners are responsible for providing and maintaining a safe workplace without limits to liability (liability is discussed in Chapter 2). Operators are required to observe applicable navigation, fisheries management, marine pollution, and safety and survival equipment regulations. However, there are presently no federal requirements for professional competency in safety, survival, navigation, seamanship, fire fighting, maintaining vessel stability, or first aid skills for over 99 percent of the uninspected fishing fleet. The Coast Guard is considering rulemaking in these areas under the Commercial Fishing Industry Vessel Safety Act of 1988 (CFIVSA).

Despite the efforts of trade associations and industry organizations, the U.S. commercial fishing industry has no effective system to promote, monitor, or require accountability of those responsible for operational and occupational safety at sea. The industry and government have relied on a combination of tradition, common sense, voluntary measures, informal guidance, marine surveys (generally for insurance), and some basic safety laws and regulations. Direct safety intervention by the federal government is coincidental with enforcement of maritime laws and regulations (particularly fisheries management regulations) and responses to emergencies. Safety performance is otherwise left to owners.

Some people in the industry have developed a sense of professional responsibility for safety. These include vessel owners or operators who have adopted safety procedures as standard practice (W. A. Adler, Massachusetts Lobstermen's Association, Inc., personal communication, 1989), trade or other industry association leaders who promote safety (Melteff, 1988; Jones, 1987; Sabella, 1987; J. Costakes, Seafood Producers Association, personal communication, 1989), and segments of the industry in which large capital investment and potentially high liability force safety programs as a protective measure. Some vessel and fleet owners have established voluntary self-inspection programs, such as annual inspections or surveys, or routine pre- or post-trip equipment or maintenance checks (see Zapata Haynie Corporation, 1989). In other cases, insurance underwriters have preconditions to issuing insurance (Nixon et al., 1987; Pacific Fisheries Consultants, 1987; McCay et al., 1989). They may require marine surveys or navigation and safety equipment checks. These requirements vary from cursory to thorough; basic checkoff lists are available from some trade associations and vendors, but data on the overall effectiveness of such programs are not available.

On a broader scale, standards of professional responsibility for safe operation have not been universally accepted and are largely nonexistent among the majority of fishing industry vessels. These include, for example, workplace procedures, regular inspections, safety meetings, training programs, protective clothing, environmental protection programs, accident investigation procedures, emergency response procedures, fire-fighting training, and—to a lesser degree—safety and survival training. Although vessel operators are required to comply with federal navigation, safety, and marine environmental protection regulations, insofar as performance can be estimated, a large number of violations in these areas are recorded annually by the Coast Guard.

Inattention to safety appears more prevalent where fisheries are depressed or where the fishermen are marginal producers. In these instances, investments in safety training and equipment may be regarded as prohibitively expensive, and fishermen may resist safety initiatives—even those that have demonstrated positive economic returns.

PRIOR SAFETY IMPROVEMENT EFFORTS

Although numbers of vessel and personnel accidents have remained consistently high in the data reviewed, safety is not a forgotten issue. Since the late 1960s, safety issues have been discussed nationally and internationally, research has been done, and programs have been developed—albeit with undetermined effectiveness. What is emerging nationally, however, is a tentative willingness within the industry to objectively consider meaningful, workable, and affordable solutions to safety problems.

Federal Safety Programs

The U.S. government has historically addressed safety in the fishing industry through the Coast Guard and the National Oceanic and Atmospheric Administration's (NOAA) National Marine Fisheries Service (NMFS) and National Sea Grant College Program (see box). Various government-supported efforts—including the present study—have considered the problem and identified possible solutions (USCG, 1983, 1971; Pizzo and Jaeger, 1974; Ecker, 1978; National Transportation Safety Board [NTSB], 1987).

Principal Coast Guard activities have focused on SAR and voluntary programs directed toward material conditions and operating procedures (Piche et al., 1987; Nixon, 1986). This does not mean that the Coast Guard has neglected human factors; however, there is greater understanding and experience treating technical systems. This is an area in which the Coast Guard participated in developing international standards for fishing vessel design and construction. There are few specific safety requirements for uninspected vessels, however, and they are primarily focused on equipment requirements.

Licenses or permits are required to operate most uninspected commercial vessels except for fishing vessels weighing less than 200 gross tons (46 U.S.C.A. §8304, 8701, 8901-8904). For documented vessels of 200 gross tons or larger operating on the high seas or otherwise subject to formal inspections, there are also requirements for officer competency and minimum manning levels for deck and engineering officers (46 U.S.C.A. §7101, 7313, 8101, 8304). Qualifications are attested to through examinations leading to a license or other documents.

In 1971 the Coast Guard completed a cost-benefit analysis of commercial fishing vessel safety programs. It considered fatalities and vessel losses for 10 selected fisheries covering 13,000 documented vessels, about two-thirds of the 1967 documented fleet (equivalent to 40 percent of the 1990 documented fleet). The agency's report found that "a full program of materiel standards, inspection, and licensing masters" would prevent 72 percent of the fatalities and 78 percent

GOVERNMENT ROLES IN FISHING VESSEL SAFETY

- **Congress** — establishment of laws and authorization and funding of government regulations and programs
- **Department of Transportation (DOT)/U.S. Coast Guard (USCG)** — marine safety, maritime law enforcement, and navigation aids
- **DOT/Maritime Administration** — marine education and training
- **National Oceanic and Atmospheric Administration (NOAA)/ National Marine Fisheries Service (NMFS)** — fisheries management
- **NOAA/National Weather Service** — marine forecasting
- **NOAA/National Sea Grant College Program** — fishing industry research and publications
- **Occupational Safety and Health Administration (OSHA)** — occupational safety and industrial hygiene
- **National Transportation Safety Board** — investigation of major marine accidents

of the property damage. However, it also found that the industry could not sustain such a program without causing many fishermen financial hardship. The report instead recommended mandatory standards for emergency equipment for documented vessels, required relatively low-cost items for preventing personal injuries, and proposed mandatory licensing of masters, to be phased in over an extended period. It also recommended a Coast Guard advisory and enforcement role that would include annual vessel compliance inspections and offer advice on maintenance and repair (USCG, 1971).

In 1974 the NMFS Northwest Regional Office studied safety and loss-prevention alternatives for Pacific Northwest and Alaska fisheries. The study—though out of date in some respects—is remarkably current concerning information gaps and causes of accidents and ways to prevent them. The study concluded that human error was an overriding cause of fishing industry accidents. It also found that:

- the data are inadequate for accident analysis;
- the industry lacks cohesiveness and continuity;
- understanding of cause-and-effect relationships relevant to accident prevention programs is poor;
- there are no explicit mandates for government programs;
- agency budgets and personnel resources are inadequate to implement effective accident reduction programs; and

- there is a strong need to alter safety-related attitudes and behavior among fishermen.

The study recommended on-site training and awareness programs as a primary method for communicating and transferring information and techniques to fishermen (Pizzo and Jaeger, 1974).

There is no effective national coordination of training and education for the fishing industry, though there are some regional efforts. Congress did not authorize or appropriate funds to implement most of the recommendations in the 1971 Coast Guard report, but discretionary funds have been available for certain Coast Guard and NOAA programs. Through them, the agencies have employed voluntary and promotional techniques to advance safety (Piche et al., 1987). These include:

- printed materials, such as safety booklets, flyers, handbooks and manuals, fishermen's digests, and newsletters;
- promotional campaigns;
- conferences (and conference reports) and safety booths at fishing trade shows;
- law enforcement boarding programs;
- voluntary design and construction guidelines; and
- cooperative ventures with industry to develop up-to-date safety manuals for the North Pacific, Gulf, and Atlantic coasts.

A Coast Guard/DOT fishing vessel safety initiative was a major voluntary safety effort. In 1986 the Coast Guard developed *Navigation and Vessel Inspection Safety Circular 5-86* (NVIC 5-86) to provide voluntary vessel design, construction, operation, lifesaving, and fire equipment guidelines. These guidelines, coupled with promotion of local education programs, were actively promoted to help the industry improve its safety record. The NMFS funded and the Coast Guard assisted in developing the North Pacific Fishing Vessel Owners' Association *Vessel Safety Handbook*, which has served as the model for a Gulf Coast handbook as well as one currently under development for the Atlantic Coast (Sabella, 1986; Hollister and Carr, eds., 1990).

A formal Coast Guard voluntary fishing vessel safety program was established with publication of a commandant's instruction charging all district commanders with a proactive district fishing vessel safety program, including assigning a safety coordinator to work with fishing interests to promote safety. The voluntary program (including NVIC 5-86) and the *Vessel Safety Handbook* were favorably received by the International Maritime Organization (IMO) Maritime Safety Committee and its technical subcommittee. U.S. delegates to IMO subcommittees actively participated in revising the technical provisions of the Torremolinos International Convention for the Safety of Fishing Vessels (1977) (discussed later in this chapter). At this time, the effectiveness of the Coast Guard's uninspected fishing vessel safety improvement activities cannot be determined, because complete data and cogent measurement criteria have not been developed.

Federal Regulation and Compliance

On the regulatory side, there are longstanding requirements for basic safety equipment (such as fire extinguishers) and lifesaving devices. These are principally enforced through underway law enforcement boardings of vessels while they are engaged in fishing. However, before the CFIVSA expanded requirements for safety and survival equipment, the Coast Guard did not have the authority to implement more stringent measures than those enabled by 46 U.S.C.A. §4102. Earlier governmental initiatives to regulate the industry were thwarted by strong resistance from the industry, which persuaded Congress not to extend enabling authorities (Piche et al., 1987). Recent efforts to seek federal relief from escalating insurance costs by amending the liability aspects of the Jones Act (Chapters 2 and 7) met similar powerful resistance, this time from the legal profession (Pacific Fisheries Consultants, 1987; Yoder, 1990). The insurance crisis of the 1980s returned fishing vessel safety issues to the congressional level. Notorious losses of fishing vessels with all personnel aboard during the mid-1980s, subsequent intense political lobbying by concerned and affected citizens, and removal of Jones Act amendments from consideration for legislative action resulted in passage of the CFIVSA, with its specific requirements for safety actions and expanded enabling authority for the Coast Guard (U.S. Congress, House, 1987, 1985; U.S. Congress, Senate, 1987, 1985; Naughton, 1990).

The federal agencies with compliance programs affecting fishermen are the Coast Guard, NMFS, and OSHA. The Coast Guard is the principal federal agency with established safety compliance programs affecting fishermen. NMFS compliance activity addresses fisheries management, but has no safety emphasis. OSHA compliance activity has primarily directed efforts toward nonfishing industrial activities aboard fish processing vessels. (Under federal regulations these technically do not include vessels that remove only head and guts prior to stowing fish product.)

Industry Safety Programs

The commercial fishing industry has promoted safety by:

- sponsoring national and regional workshops on fishing vessel safety;
- conducting research on casualty data and factors influencing safety;
- publishing newsletters and articles on vessel safety;
- conducting education and training by trade associations, local schools, community colleges and vocational training centers, and Sea Grant colleges;

A Coast Guard boarding party conducting an underway boarding of a trawler in New England waters. (PA2 Robin Ressler, *U.S. Coast Guard*)

- making safety manuals, videotapes, checklists, and similar self-help materials available through trade associations and journals; and
- conducting training programs for basic seamanship, safety and cold water survival, navigation and piloting, and similar fundamental topics, often with support from federal discretionary funds and in cooperation with federal agencies such as NOAA and the Coast Guard.

Participation in training programs on each coast and in Alaska is limited. The training sector estimates that at best, only 10 percent of active fishermen have attended at least some programs (Colucciello, 1988; NTSB, 1987). Furthermore, there is a lack of resources to support continued coordination of training and education for the fishing industry. For example, the National Council of Fishing Vessel Safety and Insurance (NCFVSI) and the National Sea Grant College Program are examining the feasibility of establishing a national network of fishing vessel safety and sea survival instructor training programs modeled after the local instructor training program used by the Alaska Marine Safety Education Association (AMSEA). The focus of this effort is on ways to develop strong regional training networks, which could be coordinated loosely within a national framework. A lack of funding precluded a research program to catalyze establishment of such a network. NCFVSI has continued data collection, and Sea Grant marine advisory programs in some regions have attempted to coordinate their efforts and share training resources to provide

Survival training in progress. Safety and survival training programs are available from a number of trade associations, academic and training organizations, and private contractors. (*North Pacific Fishing Vessel Owners' Association*)

some uniformity in curriculum and training methods (R. C. Roush, NOAA, personal communication, 1989; Melteff, 1988; Keiffer, 1984).

Also notable are efforts in local fishing communities to implement safety programs beyond those required by the CFIVSA. Some fishermen and fishing communities have banded together to provide self-help assistance during emergencies, such as in one small lobstering community in Maine (Day, 1990). There are also self-insurance groups on both the East and West coasts. While their outfitting and maintenance requirements (beyond minimum federal equipment requirements) vary, they seem capable of improving safety performance (Nixon et al., 1987; Pacific Fisheries Consultants, 1987). In some cases the fishing community has fostered development and widely adopted safety equipment before it became mandatory. A good example of this is the immersion suit, which has been widely used and is credited with saving the lives of many fishermen in northern waters (NTSB, 1987, 1989e).

International and Foreign Government Safety Activities

Various international conventions promote the safety of ships at sea. They include the Safety of Life at Sea (SOLAS) and Load Lines conventions. Most fishing vessels operating domestically are exempt from these and other conventions, but certain fish processing and tender vessels are required to have load lines (46 U.S.C.A. §5102). Recognizing the need for attention to safety of commercial fishing vessels, the IMO (formerly the International Maritime Consultative Organization, IMCO) organized an international conference, which culminated in the Torremolinos International Convention for the Safety of Fishing Vessels in 1977 (IMO, 1977). It established uniform principles and rules regarding design, construction, and equipment for fishing vessels 24 meters (79 feet) in length and over. This convention has not yet entered into full force, since not enough countries have ratified it because of concerns over technical provisions and policy issues. This includes the United States, which has not enacted legislation that would permit administration of the full range of safety measures incorporated in the convention.

Nevertheless, the Torremolinos Convention is a major milestone. It provides benchmarks for improving safety, and many fishing nations have adopted its measures into their marine safety programs. IMO is considering a protocol to the convention that would institute technological and administrative revisions to enable it to enter into force through the ratification process. The Coast Guard participated in development of the convention and is supporting development of the protocol. A major issue for the United States is whether recent law—including the CFIVSA—provides sufficient enabling authority for possible ratification.

The IMO Convention on Standards of Training, Certification and Watchkeeping for Seafarers, 1978, is another important factor. Although it specifically exempts fishing vessels, it has inspired efforts to develop personnel qualification standards. Notable among those efforts are the *Document for Guidance on Fishermen's Training and Certification* (IMO, 1988) and the *Code of Safety for Fishermen and Fishing Vessels*, Part A—Safety and health practices for skippers and crews (IMO, 1975a). Other IMO codes and guidelines include the *Code of Safety for Fishermen and Fishing Vessels, Part B—Safety and Health Requirements for the Construction and Equipment of Fishing Vessels* (IMO, 1975b) and *Voluntary Guidelines for the Design, Construction and Equipment of Small Fishing Vessels* (IMO, 1980). These standards were jointly prepared by IMO and two other United Nations subsidiary organizations, the Food and Agriculture Organization (FAO) and the International Labor Organization (ILO). They provide guidance on training and education and detailed curriculum development.

There are strong safety programs among IMO member states that include equipment standards, inspection requirements, and certification or licensing of vessel operators and crews. (International activity, Coast Guard involvement, and safety programs of selected fishing nations are summarized in Appendix C.) Rules and regulations have traditionally been developed beginning with consideration of technical systems. However, there is an increasing tendency among classification societies, legislative bodies, and international organizations to look at technical, administrative, and human elements as a total system. This is because analyses indicate that human elements directly or indirectly contribute to 70 to 90 percent of all marine accidents (Dyer-Smith and De Bievre, 1988; USCG, 1989a; T. Stallstrom, Det norske Veritas, personal communication, 1990; NRC, 1976).

Canadian studies implicate human factors in accidents and urge education and training programs to improve safety (Gray, 1986, 1987a,b,c; Canadian Coast Guard, 1987; Canada, Government of, 1988). Based on its study and national mandates, the Canadian Coast Guard (CCG) developed a new standard for protective worksuits especially for fishermen (S. J. W. Stewart, CCG, personal communication, 1990) and is sponsoring development of a liferaft for small fishing vessels.

SAFETY AS A TOTAL CONCEPT

Safety problems and solutions are too often considered individually rather than collectively. Even when a broad range of safety options is considered, actual implementation is often fragmented and uncoordinated. The CFIVSA requires that certain survival-oriented measures be implemented before all problems facing the commercial fishing industry are completely examined. Yet, safety problems arise out of a complex variety of interacting factors:

- vessels—construction, design, outfitting, navigational and operating equipment, gear type, and emergency, safety, and survival equipment;
- fishermen—professional competency (training and skills) and behavior (risk-taking attitude and responsibility for safety); and
- external forces—fisheries management, economics, and weather and sea conditions.

Individual strategies may target any or all of these risk factors, but a comprehensive program must encompass all of them as a total system. Safety options that appear to be attractive and affordable when viewed alone may offer only partial solutions, draw resources away from other options, or have unintended side effects. For example, responses from the committee's regional assessments and Coast Guard rulemaking indicate strong concern among fishermen that requirements for costly, state-of-the-art equipment such as emergency position-indicating radio beacons (EPIRBs) could result in economic hardship. Implementing these and other expensive alternatives (relative to the investment in boats and gear) could economically force some owners out of business or cause them to postpone needed maintenance (see USCG, 1971).

While each alternative needs to be measured in the context of a specific problem, it is also important to understand its contribution and cost. Such an understanding of individual elements is necessary to the creation of a package of safety alternatives that together will form a program with the greatest potential for achieving meaningful improvements. Treating safety as a total concept is also a way to distinguish between theoretically desirable goals and reasonable and attainable objectives in formulating a program of corrective action.

Establishing a Direction

Despite the congressional mandate in the CFIVSA to improve safety in the fishing industry, clear national goals and objectives have not been established by government or industry to guide development and implementation of safety programs, nor have basic questions been answered:

- What realistic level of safety is to be achieved; i.e., what are acceptable casualty and fatality rates?
- What costs—culturally, technically, and economically—are acceptable for achieving these rates? and
- What is an acceptable time frame for reaching these goals?

In 1984 the Coast Guard established a goal for its commercial vessel safety program to reduce fishing vessel casualties by 1991 by not less than 10 percent (Piche et al., 1987). The results to date are inconclusive. Special compliance examination programs targeting selected local fisheries in Alaska have demonstrated the potential for short-term, resource-intensive programs to get more vessels to carry and maintain the required safety equipment (USCG, 1988b, 1989b). Yet, so far, there are no measurable changes in the aggregate numbers of fatalities and vessel casualties.

Although the data do not show reduced numbers of fishing vessel accidents and fatalities, Coast Guard efforts do have a positive effect. For example, educational and—to a lesser degree—law enforcement actions are contributing to growing safety awareness within the industry and some local improvements. Existing efforts could be enhanced by establishing long-range objectives for which a complementary program of safety alternatives could be developed and implemented as industry and government are able to evaluate the results and bear the costs.

Safety Program Infrastructure

An effective fishing safety program will require a suitable administrative structure capable of implementing safety alternatives mandated by law or regulation. It will also need a network capable of mobilizing the affected parties (fishermen, government agencies, trade associations, and fisheries commissions and management councils) and their willingness to pay for safety services.

The only government agency with a national infrastructure capable of addressing fishing vessel issues from port to fishing grounds is the Coast Guard. It is a well-established public safety organization with administrative and technical capability to develop and administer a comprehensive safety program. Despite its extensive experience, the Coast Guard has not had the resources or enabling authority to build more technical expertise among its marine safety personnel about the safe operation and loading of fishing vessels. Specific knowledge among field personnel is generally limited to what can be learned through on-the-job training and varies by exposure to commercial fishing activity. The agency does not appear to have the budget, personnel, or authority needed to expand its role in licensing and inspection. Nevertheless, it may be the logical choice in the near term to lead and oversee (but not necessarily conduct) commercial fishing safety programs.

Structurally, the Coast Guard's commercial vessel safety program resembles a total concept approach, though a full benefit-cost analysis of proposed *and* existing safety measures is not normally an element of rulemaking. When first directed by Congress to look at fishing vessel safety in 1968, the Coast Guard tried to analyze safety issues as a total concept (USCG, 1971). In practice, however, lack of congressional and industry support resulted in an incremental approach.

There is an informal network to mobilize constituent support for selected issues among the NCFVSI and some trade associations. The council and trade associations have developed safety materials, sponsored workshops, conducted safety training and research, and represented the industry during congressional hearings on vessel safety and insurance. The council consists of representatives from the major fishery trade associations across the United States. It could possibly be used to administer a comprehensive national safety program, but is not organized to do so. There is no single organization with the breadth of constituents or cohesiveness to represent the entire industry.

Although significant differences among various fishery interests and trade associations may preclude leadership by a single industry representative, the Atlantic, Gulf, and Pacific coast states' regional fishery commissions might fill such a role. Although the commissions evolved independently and thus respond differently to their regions' needs, their potential to lead safety programs regionally merits consideration. For example, the Pacific States Marine Fisheries Commission cosponsored the West Coast regional assessment commissioned for this study. Its neutral role provided a forum for obtaining the information needed to understand safety issues there.

Monitoring and Evaluating Safety Performance

Monitoring and evaluating safety equipment, systems, and programs are fundamental. A total safety system must include ways to identify safety trends, anticipate problems, and assess program effectiveness in mitigating the severity and costs of accidents. A complete system would include reporting, data collection, and analysis of criteria on which safety performance will be based.

Data systems are needed to provide for follow-up and compliance when safety performance is unacceptable.

There is presently no universal program for evaluating safety in the commercial fishing industry. Monitoring techniques tend to be rudimentary: for example, monitoring personnel injury incidents and costs or correlating accidents with diesel fuel consumption to approximate the relative effectiveness of safety actions within comparable corporate fleets. Computer software for monitoring vessel performance was introduced by the NCFVSI in 1989, but information from such programs belongs to the user and is not tracked. Insurance claims are a natural resource for casualty statistical analyses, and the CFIVSA requires insurers to provide casualty data to the Coast Guard. This is not yet being done. Some claims data are provided voluntarily by marine underwriters to the Commercial Fishing Claims Register (CFCR) maintained by the Marine Index Bureau (see Appendix D), but the data are far from complete and cover only a portion of vessels and personnel casualties, and there is no effective monitoring system.

Coast Guard data include documentation of federal marine casualty investigations, SAR services, underway and occasional dockside boardings, and courtesy examinations. Unfortunately, the data record only limited safety performance information across the entire fleet of uninspected fishing vessels. Prevalent low-level maintenance deficiencies are indicated, and a close estimate of the number of annual fatalities is supported by the data. Secondary analysis of causal factors was only possible for about 30 percent of the fishing vessel casualties recorded in CASMAIN. Directly correlating information among data bases was not possible, however.

SUMMARY

Safety problems abound within the commercial fishing industry, despite past efforts by government and industry to correct conditions. These efforts for the most part have approached the complexities of safety improvement in a fragmented and uncoordinated way, rather than as a total concept. Goals and objectives needed to establish an overall safety program have been lacking. Safety trends cannot be effectively monitored and evaluated using present means, and without accountability, fishermen have been reluctant to accept a proactive role in safety improvement. The CFIVSA signals that this is an opportune time to address safety as a total concept leading to adoption of a comprehensive safety program involving cooperative ventures by government and industry.

In the following chapters, attention is focused on the broad safety areas identified in this chapter—safety performance, the vessels, the fishermen, survival, and external forces. Specific safety issues are discussed and alternatives considered. Specific problems and possible solutions are addressed as discrete components that can be fitted into a safety program with a national infrastructure and a means of evaluating it.

2
Commercial Fishing: An Industry Overview

The United States—with coastline on two oceans, the Gulf of Mexico, the arctic seas, and the Great Lakes, and its extensive rivers, lakes, and reservoirs—is among the leading fishing nations of the world. Fishery resources within its 200-nautical-mile exclusive economic zone (EEZ) make up about 15 percent of the world's total. Commercial fishing makes significant contributions to the national and regional economies; in 1989 10.7 billion pounds of fish were landed by U.S. vessels, fifth in total world harvest behind Japan, the Soviet Union, China, and Peru (National Oceanic and Atmospheric Administration [NOAA], 1990a).

The U.S. fishing industry is composed of harvesting, processing, and marketing segments, each with an associated infrastructure. The numbers of vessels used in this report are estimates based on composite data of widely varying statistical validity and are presented to provide a frame of reference for development and analysis of safety-improvement strategies and alternatives later in this report. About 30,000 fishing industry vessels were documented with the federal government in early 1990 (vessels 5 net tons and over). Table 2-1 depicts the actual number of documented vessels that could be categorized as "fishing industry vessels" on March 31, 1990. The ports in which these vessels are documented do not necessarily reflect the regions where they are employed. For example, a significant number of vessels from the West Coast, and to a lesser degree from the North Atlantic region, are operated in North Pacific and Alaskan waters. Fish processing and fish tender vessels are operated almost exclusively in North Pacific and Alaskan waters (a few operate in the North Atlantic region). An estimated 260 fishing industry vessels have federal

TABLE 2-1 Documented Self-Propelled Vessels Under 5,000 Gross Tons with a Fisheries Endorsement but not a Passenger or Offshore Supply Vessel on March 31, 1990

Coast Guard Documentation Port	Number of Vessels
Atlantic Coast	
Boston, Massachusetts	3,255
New York, New York	950
Philadelphia, Pennsylvania	715
Hampton Roads, Virginia	3,668
Miami, Florida	3,467
Great Lakes	
Cleveland, Ohio	155
Gulf Coast	
New Orleans, Louisiana	3,264
St. Louis, Missouri	55
Houston, Texas	2,224
West Coast	
Long Beach, California	974
San Francisco, California	1,945
Portland, Oregon	1,522
Seattle, Washington	2,835
Alaska	
Juneau, Alaska	4,335
Hawaii/Southwest Pacific	
Honolulu, Hawaii	305
Total	29,669

SOURCE: Data recorded in U.S. Coast Guard Marine Safety Information System by Coast Guard Headquarters, Washington, D.C., on March 31, 1990.

or state permits to process (e.g., freeze or can) fish. Of these, about 210 have both harvesting and processing capabilities. A documented vessel's actual use in the fishing industry is not monitored by Coast Guard automated information systems or data bases.

In 1987, the latest year for which broad-based industry data are available regionally, it is estimated that about 31,000 federally documented fishing industry vessels and 80,000 smaller craft were registered with the coastal states (with the Coast Guard in Alaska) and bearing state numbers (Table 2-2). These

TABLE 2-2 Estimated Number of Fishing Industry Vessels Active During 1987 by Region Fished[1]

Region	Documented Vessels	State-Numbered Vessels
North Atlantic		
New England	1,800	16,500
Mid-Atlantic	800	5,500
Chesapeake Bay	2,500	3,500+[2]
South Atlantic	2,700	13,500
Gulf/Caribbean		
Gulf Coast	10,000	26,500[3]
Caribbean	[4]	1,500
Great Lakes	[5]	[5]
West Coast	5,000	6,000
Alaska	8,000	9,000
Hawaii/Southwest Pacific	200	200
Total	31,000+/-	80,000+/-[6]

[1]Numbers are composite estimates from regional sources. Principal sources include records of fish landings maintained by National Marine Fisheries Service regional offices, permit data maintained by the Commercial Fishing Entry Commission in Juneau, Alaska, and regional assessments commissioned for this study, and economic analyses available for some fisheries.
[2]Based on 1986 estimate of Chesapeake Bay oyster fishery (Sutinen, 1986).
[3]Includes a large number of small boats engaged in shrimp fisheries in bays, sounds, and estuaries.
[4]Negligible.
[5]Current information is not available.
[6]The number of commercial fishing vessels bearing state numbers is not known. West Coast and Alaska figures are close approximations. All other data presented are general estimates.

vessels and small craft employed many types of gear to catch, transport, or process a wide variety of finfish and shellfish. The number of vessels constructed as fishing vessels but not actively used in fishing is not known.

The number of individuals who fish commercially is not known, nor is there a statistically valid average number of fishermen per vessel. The number of fishermen varies from 1 to 20 or more, depending on the size of the vessel and its fishing activity. Generally, the majority of the documented fleet is estimated to have three to four fishermen per vessel, and state-numbered vessels one to two. If all the vessels were under way at the same time, this would equate to a capacity for about 230,000 jobs. However, the actual number of individuals is probably significantly higher. This is because many people are hired as part-time and seasonal workers, and there is high turnover among entry-level crewmen. Thus, the number of people employed as fishermen that is used in this report is only a crude estimate to provide a reasonable frame of reference.

The U.S. catch landed at U.S. ports in 1989 totaled nearly 8.5 billion pounds (3.8 million metric tons), with a value of $3.2 billion. Commercial landings by U.S. fishermen at ports outside the 50 states or transferred to foreign vessels within the U.S. EEZ were an additional 2.2 billion pounds (994,000 metric tons) valued at $326.7 million. Most of this consisted of tunas landed at canneries in American Samoa, Puerto Rico, and foreign ports and pollack transferred to foreign processing ships within the U.S. EEZ.

Within the processing and wholesale sector, there are about 4,200 establishments employing 103,000 people (annual average). These operations process and market fishery products throughout the United States and abroad. Some processing is conducted aboard ship, principally in North Pacific and Alaskan waters. In addition, U.S. processors import 3.2 billion pounds of edible seafood products valued at $9.6 billion for further processing before they are marketed. The U.S. commercial marine fishing industry contributed $17.0 billion (in value added) to the U.S. gross national product in 1988 (NOAA, 1989) and $17.2 billion in 1989 (NOAA, 1990a).

Imported seafoods are becoming an increasingly important source of products for America's seafood consumers. Although imports have continuously represented less than 50 percent of the total edible seafood supplies in the United States, since 1966 they have increased at an annual average rate of about 5 percent. In contrast, domestic landings have increased at about 2 percent. If menhaden (used for industrial products) and Alaskan pollack are excluded, there has been a decline.

International trade in seafood has become a dominant force that shapes the economic performance of the commercial fishing industry. If the growth of the U.S. industry is constrained by external factors affecting fishery resource management or declining resources force curtailments, imports will fill the demand. As some countries rapidly expand aquaculture production, prices for wild caught species, especially shrimp and salmon, may be undercut (McDowell et al., 1989). Although it is beyond the scope of this study, the committee notes that erosion of economic returns from U.S. fisheries competing with aquaculture products may increase economic pressures in some fisheries, detracting attention from safety.

FISHING INDUSTRY VESSELS AND FISHERMEN

This introduction to the commercial fishing industry turns now to the vessels, the people who earn their living aboard them, and the working conditions in which they pursue their profession. These are the areas where safety problems occur, which motivated the Commercial Fishing Industry Vessel Safety Act of 1988 (CFIVSA) and this study. The U.S. commercial fishing fleet is one of the world's largest in total numbers in which most of the craft are linked to coastal and estuarine fish stocks. Since the 1976 Magnuson Fisheries Conservation and Management Act (MFCMA) was implemented, the U.S. fleet has expanded and modernized to a large degree. Yet, many old vessels remain in the fisheries and will, under present circumstances, continue to be operated for many years. The fishermen who operate these vessels are as diverse as the vessels themselves.

The Vessels

In simplest terms, commercial fishing vessels are self-propelled or wind-driven platforms used to catch fish for profit. In the broadest sense they are a workplace, a means of transportation to and from the fishing grounds, an itinerant domicile for overnight or extended trips, and for some an industrial plant for processing products. Although it is correct to refer to them as "vessels," many can be appropriately considered "boats," reflecting their relatively small size. In length, vessels of the U.S. fishing industry fleet range from under 25 feet to over 300 feet. But the majority of the fleet are small fishing vessels; about 99 percent are 79 feet or less in length. It is estimated that roughly 80 percent are less than 40 feet. Their hulls are of wood, aluminum, steel, fiberglass, and even concrete. In age, they range from those under construction to those constructed prior to the turn of the century. Their fishing riggings include various types of nets, trolling gear, trawls, hooks, dredges, rakes, and traps.

Vessel Technology

Tremendous progress has been made in new vessel design and construction. Shipbuilding techniques and new fishing experiences are reflected in the design, construction, and operations of modern fishing vessels. Hydraulic power is used on many vessels to operate most deck gear through remote workstations. Electronic navigation, communications, and fish-locating equipment fills many bridges. Mechanical cooling and freezing equipment is common, particularly on vessels operating on longer trips. Factory trawlers and ships carry industrial processing equipment as well. Midwater and off-bottom trawls, techniques employed in European fisheries for many years, are now successfully employed in U.S. fisheries of the Pacific and Atlantic oceans and the Gulf of Mexico.

What has emerged is a modern, efficient fleet capable of taking large quantities of fish and successfully replacing the foreign fleets in the offshore fisheries. Many traditional fishing boats and small craft continue to dominate the coastal, estuarine, and Great Lakes fisheries. These smaller craft employ fishing gear much like that used in the past. The major exception has been the menhaden fishery, where vessels, gear, and efficiency rival that of the modern offshore fleet.

Commercial Fishermen

The men and women who work the fisheries of the United States are as diverse as the fish they seek. They are of all races and ages. Some are first-generation fishermen; some come from generations of fishermen, often with one or more family-owned boats. Some have no alternative marketable skill or source of employment and fish as a source of income or subsistence. Others become commercial fishermen because they like it and they have a share in the vessel's earnings (see Nixon, 1986). Therein lies a major difference from the fisherman's onshore counterparts. As a rule, fishermen get no guaranteed wage, no overtime pay, and few fringe benefits. They get only the promise of hard work, long hours, a high-risk workplace, and—by any standards—cramped living quarters in exchange for a share of the net proceeds at the end of the trip.

A fisherman may be a fishing vessel owner who serves as captain of his own vessel (owner/operator), a person employed by the owner to operate the vessel (operator, individual in charge, or captain), or a crewman. Used collectively in this study, the term fisherman applies to the captain and all members of the crew engaged in service on deck or in engineering departments or capacities aboard a fishing industry vessel. Principal occupational activities include vessel, fishing, harvesting, and delivery operations. Processing-line workers aboard catcher/processors or floating processors are also part of a vessel's crew. But they are not employed in the occupational activities associated with fishermen, instead performing an industrial function.

Functionally, processing-line work is best characterized as unskilled. Entry-level line workers frequently have extremely limited, if any, maritime experience. Some speak English only as a second language, a complicating factor during emergencies. Although they are not characterized as fishermen for this study, their safety is an issue insofar as they may be jeopardized by vessel operations, and the analysis of safety presented in this report applies. Processing-line safety is an issue of concern but is beyond the scope of this study.

Culture and Social Organization

Fishermen are often viewed by social observers as a quaint subcultural group displaying special social and cultural qualities: individualistic, carefree, rugged, self-sufficient, and in some cases fatalistic. Fishermen more frequently characterize themselves as hunters. During this study, the committee was continually struck by what appear to be basic social and psychological assumptions (sometimes verging on stereotypes) of fishermen as possessing social

The lonely vigil—a West Coast albacore fisherman waiting for a strike. (*Oregon Sea Grant*)

characteristics far from the mainstream culture, similar to earlier sociological conceptualizations of farmers (Gross, 1958). This may be the case in some fishing communities, particularly those where fishing is a significant element of the local economy. Also, demands of the workplace distinguish fishermen as a unique occupational group (Browning, 1980; Maiolo, 1990)—e.g., their temperament and the ability to endure long periods of boredom and isolation and long hours of physical labor (Browning, 1980).

The fishing industry has attracted people from many ethnic groups. Some are immigrants who speak their native language and may have little understanding of English. Most quickly assimilate into the workplace with minimal problems. However, as technology has made fishing operations more complex, familiarity with a common language is a necessity. During the regional investigations, anecdotal information indicated that language barriers may hinder communication and contribute to accidents. This is mostly a regional factor that is typically associated with relatively few ports, areas, and fisheries (e.g., shrimp and tuna) and particular operations (e.g., factory trawlers).

Fishermen can be classified in terms of their occupation as skilled. Perhaps the major factor distinguishing them from other occupations is the eclectic nature of the skills required, ranging from shipwright and diesel mechanic to refrigeration engineer and electronics technician. The level of knowledge required varies significantly by vessel, gear configuration, and status as captain or crew member. As with other occupations, fishermen are differentiated by skills that distinguish the successful from their marginal counterparts. For most of the fleet, which is small-scale in nature, the role of the fisherman is "similar in many respects to that of the independent farmer, the one-business merchant, the autonomous professional" (Miller and Van Maanen, 1982).

Fishermen as Participants in the Labor Market

An alternative view of fishermen as a segment of the labor market was prepared for this study (Gale, 1990). Analysis of labor market behavior of fishermen suggests the general approach they take with regard to the workplace, where accidents occur. Accidents, probabilities, and reactions to vessel inspection and other possible safety programs would be influenced by both work setting characteristics (e.g., gear type) and fishermen's behavior and motivation.

Gale suggests there are different generic characteristics of how fishermen approach fishing as an occupation, and this affects how they respond to safety considerations. There are those who have a strong commitment to fishing, some of whom will adapt to change and others who will resist it. Some fishermen enter and leave fishing in order to supplement their income with other work. Those who are not primarily fishermen pursue the short term monetary reward. Each responds to safety considerations differently. Thus, fishermen differ in their response to the labor market, their commitment to the occupation of fishing (in contrast to specific tasks or jobs), their motivation for making fishing a career, and their reaction to safety and safety programs. Gale suggests that no single strategy will solve all safety problems and that rigid programs will vary in effectiveness depending on the relationship of the targeted fishermen to fishing as an occupation.

As regulations mandated by the CFIVSA and safety decisions deriving from it come into effect, fishermen—people who historically have lived with risk and danger, who often display characteristics of subcultures, and who display intense individualism—will be compelled to meet safety standards mandated by a distant decision-making body, the Congress of the United States. A chief implementation problem may be cultural and social uniqueness; however, commercial fishermen in the United States have found ways to cope with previous mandates, such as the resource management schemes of the MFCMA and ensuing state and regional management developments. The indicators of adjustment are increased involvement of commercial fishermen in the management process, such as participation on advisory panels, attendance at hearings, the strengthening of association ties to promote their interests, and the use of political influence. Such adjustments have not always come easily in the past, but what appears important is bringing those who must comply with new regulations into the decision-making and implementation processes as participants.

The Working Conditions

Working conditions throughout the national fleet and the fishing grounds further complicate safety. R. M. Snyder (1973), an early oceanographer, succinctly put into context the ocean as a working environment:

> There are things about the sea which man can never know and can never change. Those who describe the sea as "angry" or "mean" or "gentle" or "ferocious" do not know the sea. The sea just doesn't know you're there—you take it as you find it, or it takes you.

There is no room for carelessness or arrogance on the ocean. Fishermen are continuously exposed to high risk in the workplace during transit and while fishing. They are required to work extremely long, unregulated hours, often under very severe environmental conditions. In most cases, fishermen are not required to have their professional competency validated by third parties or their physical condition screened prior to employment. In the majority of the fleet, fishing vessels are not surveyed or inspected during construction or operation to ensure satisfactory material condition (vessels do not maintain themselves). Fishing vessels obtain cargo from the sea and must be loaded at sea, resulting in open hatches and considerable deck activity. This often occurs in changing—often marginal—weather on moving vessels with high pitch and roll movement and relatively low freeboard (see Browning, 1980; Canadian Coast Guard, 1987; Murray, 1962). Furthermore, fishermen (and processing workers) for the most part are a nonunion work force. There are no third parties monitoring work hours, health benefits, time at sea, profit sharing (most fishermen do not work for wages), grievances, or collective bargaining. These working conditions are not inviolable; some elements hold potential for improving safety.

U.S. LAWS AFFECTING COMMERCIAL FISHING

There are a number of laws that have an impact on the commercial fishing industry. The principal ones important to this study are those pertaining to fisheries management and legal liability.

Fisheries Management

Traditionally, commercial fisheries have been highly competitive. Return to capital and labor has been based on sharing of revenue generated by sale of catches, either by the trip or over the whole season. Even in colonial days, competition was keen. The drive to be the first to market with catches from the Grand Banks is reflected in historical records. This longstanding

A large halibut hauled aboard with gaff hooks. Fishing industry vessels load cargo at sea rather than in port, creating unique operating conditions and risks to vessels and fishermen. (Robert Jacobson, *Alaska Sea Grant*)

practice of high pay for successful fishing varies little across geographic area and fisheries. Fishermen share a percentage of the proceeds from the sale of the catch after deductions for vessel and trip expenses (e.g., fuel costs, groceries, bait, electronics, leasing fees, etc.), owner's share, and bonuses or premium for skippers and other personnel. This traditional payment method produces strong economic incentives for maximizing catches and minimizing costs. These practices predate any significant efforts at fishery management, which today frequently conflict with the fishermen's desire to increase their incomes through larger catches. Presently, conservation efforts have led to quota management, resulting in shorter fishing periods.

Prior to enactment of the MFCMA, most marine fishery management regimes were within the purview of the states. Federal management was the result of international treaty obligations covering tunas, Pacific halibut, West Coast salmon, and most finfish and some shellfish of the northeast Atlantic. These earlier conservation regimes addressed seasons, gear limitations, national quotas, and some area closures and size limits. Overfishing of stocks was attributed to foreign fishing activity, and efforts to reduce fishing effort focused on foreign fleets. Thus, recent regulatory regimes to control direct fishing effort have their beginnings in the international forums or (since 1976) within the eight fishery management councils created by the MFCMA (Grasselli and O'Hara, 1983; Nies, 1986; Sutinen and Hennessey, 1986; Alverson, 1985).

The MFCMA authorized the federal government to conserve and manage all fishery resources except tuna within the U.S. fishery conservation zone (200 nautical miles from the coast). The eight regional fishery management councils (FMCs) have the principal responsibility for developing fishery management plans (FMPs), which generate management measures for particular fisheries. Fishing industry views are represented to the regional councils through industry advisory committees and opportunities to speak at council meetings open to the public. Inasmuch as most fisheries are specific to a geographic area, fishermen transferring from one region to another are confronted with a wide variety of regulations (Grasselli and O'Hara, 1983; Nies, 1986; Sutinen and Hennessey, 1986; Frady, 1985).

The FMCs have continued the traditional management practices of the past. During this same period, many U.S. fisheries have witnessed significant technological improvements in vessels, deck gear, electronic fish-finding and navigational equipment, and fishing gear and technology. As a result, many stocks of finfish and shellfish have declined, in large part from overexploitation by U.S. fishermen. This has resulted in more restrictive conservation regimes, designed to arrest declines, begin stock rebuilding, or preclude overfishing (Frady, 1985). Whatever the reason, whichever the fishery, the restrictive measures—coupled with greater fishing effort—have resulted in economic pressures on fishermen with indirect adverse effects on safety.

Legal Liability

The Jones Act (46 U.S.C. §883) requires that the coastwise transportation of merchandise (i.e., goods, wares, and chattels of every description including fish and fish products [19 U.S.C §1401]) between two points in the United States must be on board a U.S.-built and -flagged vessel. Beyond that, the act is best known for establishing an injured seaman's right to trial by jury for any action for damages suffered at the hand of his employer. This, together with the general maritime doctrine of unseaworthiness that enables all persons in service of the vessel to recover full indemnity if injury was caused by any condition of the vessel, its equipment, or crew that a court may construe as rendering the vessel "unseaworthy," is a basis for liability findings and settlements. The system effectively incorporates the potential for high awards inherent with fault-based liability and the virtual certainty of recovery usually associated with no-fault or workmen's compensation types of systems. This legal liability regime is considered onerous and economically disruptive by many fishing vessel owners, because it generates problems in the pricing and availability of liability insurance. (Chapter 7 discusses the impact of this legislation further.)

REGIONAL FISHERIES

The commercial fishing industry has traditionally been regional in character. Unique regional factors include fishery stocks, environmental conditions, and significant interregional issues that affect safety, such as fishing outside the vessel's or fisherman's home region and converting vessels designed and constructed for one region or fishery for use in another. The regional nature of U.S. fisheries was reinforced by organizational changes to fisheries management practices, mandated by Congress in 1976, which created the federal regional FMCs.

Safety analysis begins with a summary of the regional fisheries, including intra- and interregional issues and estimated 1987 baseline numbers of vessels and fishermen developed for this study (Table 2-2). Fisheries data are generally collected regionally. Timely data on vessels and fishermen are woefully incomplete, however, and the numbers are at best composite estimates based on correlation of data from national and regional sources. The number of vessels in the national fishing fleet is constantly changing. The number of fishing industry vessels active in 1987 is believed to represent a decline of undetermined size from the fleet size in 1982. A 25 percent decline is evidenced for the West Coast, for which statistically valid historical data were available on the number of active fishing industry vessels (Jacobson et al., 1990).

North Atlantic Region

The North Atlantic (New England through mid-Atlantic ocean waters and Chesapeake Bay) fishing fleet operates from Maine through Virginia in both nearshore and distant waters. Exclusive of vessels home-ported outside the region, there are about 5,100 documented fishing industry vessels operating from ports in the North Atlantic region. Of these, fewer than 12 have combined catching and processing capabilities, and there are a few fish tenders (Griffen, 1989a,b; Platt, 1990). The remainder are fishing vessels. In addition, there are approximately 25,500 state-numbered fishing vessels. The majority of the fleet operates in coastal and estuarine waters south of the Canadian border, but some vessels venture to the outer edge of the Grand Banks (outside Canadian jurisdiction) for groundfish (e.g., cod, haddock, and flounder), and others operate well offshore for swordfish and tuna. Some catcher/processors from the region have operated in the Gulf of Mexico for squid and butterfish, and at least one shifted to the North Pacific trawl fisheries during this study. There is a major oyster fishery in the Chesapeake Bay, principally manned by 3,500-4,500 fishermen in small boats employing hand or mechanical tongs. About 30 traditional sail dredges have also been used in this fishery (Sutinen, 1986).

Environmental conditions can be severe, especially during the winter months. During periods of better weather, many vessels from outside the region—some as far away as Texas—travel to these waters to fish. Twelve of the nation's top 60 fishing ports are located in the North Atlantic region, accounting for approximately 20 percent of the nation's catch by value and 9 percent by volume in 1989. Groundfish, lobster, and scallops collectively make up about 60 percent by value of the region's total. The rest is divided among other species (butterfish, clams, crab, herring, mackerel, menhaden, northern shrimp, squid, swordfish, and tunas) (NOAA, 1989, 1990a). Each of these fisheries presents its own safety problems because of the wide variety of gear and vessel types used (Allen and Nixon, 1986; Gordon, 1989; Nixon, 1990).

South Atlantic Region

The South Atlantic region extends from North Carolina to the Florida Keys. It accounted for about 5 percent of the nation's catch by value and 3 percent by volume in 1989. Fishing provides significant employment to the regional economy. About 2,700 documented and 13,500 state-numbered vessels are used to fish commercially (Jones and Maiolo, 1990; NOAA, 1990b; National Marine Fisheries Service [NMFS], unpublished data, 1990). No fish processing vessels operate in the region. There are few offshore operations in the region. Most fishing craft operate within state jurisdiction (within 3 nautical miles of the shoreline and within bays and inlets).

The region is characterized by diverse fisheries, the most important being menhaden, shrimp, crab, flounder, and calico scallop. Major inshore species are crabs and various food fish taken by traps, pots, gill nets, and seines. Slightly farther offshore (outside 3 nautical miles), a variety of finfish are taken by seiners, trawlers, trollers, and gill-netters (Gordon, 1989; Jones and Maiolo, 1990; Allen and Nixon, 1986; NOAA, 1990b).

Since nearly all fishermen depend on daily or weekly sales of fresh fish, the economic incentive to operate year round is high. Environmental conditions are generally not a limiting factor except for hurricanes, periodic storms, and sea conditions that impair departure and entry to some fishing ports. Many offshore operators, particularly those from North Carolina, also fish in the North Atlantic region, landing their catch in ports adjacent to the fishing grounds. Such interregional activity, also found on the West Coast and Alaska, greatly increases the local knowledge needed by vessel captains to operate safely.

The Gulf of Mexico and Caribbean Region

The Gulf Coast extends from the Florida Keys west to the U.S.-Mexican border. There are approximately 10,000 documented fishing vessels. No fish processing vessels are based in the region, although one vessel is being converted to process multiple species in the Gulf of Mexico. The states accounted

The fishing vessel St. Paul III typifies the vessels which operate in the Gulf of Mexico shrimp fishery. (Randall G. Prophet, *Florida Seafood Marketing*)

for 20 percent of the nation's catch by value and 21 percent by volume in 1989. Additionally, there are many small craft (estimated at 26,500)—usually one-person operations—fishing commercially in sheltered inland waters, principally for shrimp (NOAA, 1990b; NMFS, unpublished data, 1990; Hollin, 1990). Environmental conditions affecting the Gulf Coast are relatively benign. Fishermen must, however, be alert to occasional tropical storms and hurricanes as well as squalls and thunderstorms, which can occur throughout the year.

The Gulf Coast includes high-volume menhaden and shrimp fisheries. The fleet is numerically dominated by smaller craft, yet larger offshore shrimp vessels number about 7,500. Recently, there has been a move to develop a longline fleet for pelagic (open sea) fisheries. Some new vessels have been constructed for this fishery; others are older vessels converted from shrimping. Overall, the majority of fishing vessels operate within 3 nautical miles of shore. The major commercial fisheries are shrimp, menhaden, spiny lobster, stone crab, swordfish, reef fish, and oysters (NOAA, 1990b). These fisheries require diverse vessels and gear, with different operating procedures for safe handling (Gordon, 1989; Hollin, 1990; Marine Advisory Service, 1986).

The Caribbean encompasses the fisheries of the Commonwealth of Puerto Rico and the U.S. Virgin Islands. The island-based craft are small and total about 1,500. The number of fishermen is small—about 2,000. Their landings are minor in relation to those of the overall Gulf and Caribbean region and the nation. Nevertheless, they are important contributors to island economies. Ponce and Mayaguez, Puerto Rico, are important offloading ports for tuna, which is processed and canned there (Jones, 1990).

Great Lakes and Inland Fisheries

The Great Lakes (included in the scope of this study) and vast inland waterways of the United States support varied commercial fisheries. Hundreds of small, state-numbered boats, and in the Great Lakes a small number of documented vessels as well, utilize gill nets, trot lines, seines, traps, hand lines, fyke nets, trammel nets, trawls, and other minor gear. Most of the smaller boats are operated by the owners. Larger vessels operating on the Great Lakes may carry a crew of up to five. Most, if not all, vessels are operated as day boats, leaving port in early morning and returning in late afternoon. No current information is available on the number of vessels actively employed in commercial fishing. NMFS figures for 1977, the most recent vessel employment information available for Great Lakes and inland fisheries, estimated that 225 documented vessels and 586 state-numbered boats operated on the Great Lakes (NMFS, unpublished data, 1990). Commercial fishing activity has declined since 1977. Coast Guard documentation data for March 1990 and vessel employment in other regions suggest that the number of active documented vessels with fisheries endorsements on the Great Lakes is less than 150. The number of commercial fishermen on the Great Lakes is small—2,000 to 3,000. Although outside the scope of safety analysis presented in this report, the 1977 NMFS data estimated that 8,613 state-numbered boats operated in the 20 states comprising Mississippi River fisheries (NMFS, unpublished data, 1990).

Landings taken by Great Lakes and inland fishermen are small in comparison to U.S. fisheries in other regions. Nevertheless, the fisheries are important contributors to local economies and frequently are the major source of fishery products on local markets.

West Coast Region

There are nearly 5,000 documented fishing vessels operating in waters off the West Coast, providing work for about 11,000 fishermen. An estimated 6,000 fishing vessels bearing state numbers also operate in these waters (Jacobson et al., 1990). The number of fishermen operating state-numbered vessels is not known. Fleet profiles developed for Washington State waters estimate that 3,525 fishing industry vessels employed 6,972 fishermen during 1987 (Natural Resource Consultants, 1988).

West Coast states (California, Oregon, and Washington) accounted for about 10 percent of the nation's catch by value and 9 percent by volume in

1989. From northern California to Washington, the salmon fishery is the most valuable. It also employs the greatest number of gill net, troll, and seine vessels. The trollers are relatively small (18-60 feet), yet operate up to 50 nautical miles offshore throughout the salmon and tuna fisheries. Gill-netters and seiners operate in the relatively protected waters of estuaries, bays, and inlets during the seasonal salmon fishery. During 1987, about 85 percent of vessels operating in Washington State local waters were engaged in the salmon fishery (Natural Resource Consultants, 1988). No U.S.-flag fish processing vessels were identified as operating in West Coast waters. There are limited fish tender operations, consisting principally of fishing vessels seasonally converted to tendering for river runs of salmon in sheltered inland waters (Jacobson et al., 1990).

Fishing vessels throughout the region, especially from northern California through Washington, face a full seasonal range of environmental conditions offshore, including high winds and seas and fog. There are few areas along the entire coast outside of harbors in which to seek shelter. The major fishing ports from northern California to Washington are renowned for hazardous local conditions. Vessels frequently must ride out storms at sea or risk crossing hazardous bars to enter port. Coast Guard search and rescue (SAR) data disclose heavy "clustering" of emergency assistance incidents in these areas. Additionally, the coastal waters of northern California and Oregon are frequently fogbound and are heavily used by coastal freighters and fishing vessels in transit, increasing the threat of collision (see Bard, 1990).

Distant-Water Fleet

There are major fisheries based in the United States that operate both in and beyond the U.S. EEZ, such as the vessels from the West Coast and Hawaii that engage in distant-water fisheries. The distant waters include the U.S. possessions or trust territories of the central and western Pacific. They also include North Pacific and Alaskan waters for a large number of vessels home-ported on the West Coast.

The major fishing grounds for many vessels from Oregon and Washington are in the Bering Sea, Gulf of Alaska, and Alaska State waters for salmon, king and tanner crabs, halibut, and groundfish. The number of vessels engaged in these fisheries is included in the estimate for the Alaska region. Fleet profiles developed for vessels operating from Washington State but fishing in the Alaska region estimate that 1,552 fishing industry vessels employed 8,163 fishermen during 1988. About 70 percent were engaged in the salmon fishery. There were also about 43 factory trawlers and 25 mobile crab processors (Coopers and Lybrand, 1990; Natural Resource Consultants, 1988).

Prior to 1976, the groundfish fishery off Alaska in what is now the U.S. EEZ was dominated by foreign fishing vessels. "Americanization" of this fishery has taken place, initially through U.S.-foreign joint ventures, and subsequently through a combination of shoreside processing operations and catcher/processor vessels. Joint ventures with foreign-flagged vessels are being phased out (although there are significant foreign investments in some U.S.-flag processors and catcher/processors), and new joint venture arrangements are occurring domestically through alliances between U.S.-flag factory processing and fishing vessels. Many of the larger vessels are based in Puget Sound and other Pacific Northwest ports (Coopers and Lybrand, 1990; Natural Resource Consultants, 1986, 1988; Campbell, 1990). There is some acrimony between Alaska operators and fishing operations based in Oregon and Washington competing for the available resources, and there is some evidence that the fishery may be overcapitalized (Bernton, 1990; Campbell, 1990; Wiese, 1990; Matsen, 1989). It was estimated that during 1990 the fleet would grow to 54 factory trawlers and 3 mother ships, with additional modest growth anticipated (Coopers and Lybrand, 1990; Natural Resource Consultants, 1988, 1989; Munro, 1990). In July 1990, Alaska Factory Trawler Association (now American Factory Trawler Association) records indicate the number had grown to 68 active factory trawlers.

Most notable of those fishing outside the EEZ are the tuna fishing vessels based in San Diego and Los Angeles, California, or Honolulu, Hawaii. The U.S. tuna industry today operates tuna fishing fleets virtually throughout the tropical oceans. Principal fishing areas are in the eastern, central, and western Pacific, and some in the Atlantic and Indian oceans. Targeted species are yellowfin,

A modern U.S. flag factory trawler. (David Sears, *American Factory Trawler Association*)

A loaded cod end being brought aboard a factory trawler configured with a stern ramp. Industrial safety equipment such as hard hats are in use aboard some vessels. (*American Factory Trawler Association*)

skipjack, and albacore. Bigeye, bluefin, bonita, and black skipjack are also taken. Abut 65 large superseiners operate throughout the tropical Pacific, some venturing into the Atlantic and Indian oceans pursuing yellowfin and skipjack tuna. Somc distant-water vessels operate off Central America, and some venture farther into the south-central Pacific. Most of these are based in U.S. ports and utilize fishing grounds within the fishery zones of one or more nations. Smaller vessels (about 165) operating from Hawaii and the West Coast use longline or troll gear to fish for bigeye, bluefin, and albacore (Bourke, 1990).

Alaska Region

If Alaska were an independent country, it would probably rank near the top 10 nations of the world for fisheries production by volume and value—in 1986 it would have ranked number 11 (McDowell Group, 1989). It accounts for nearly 50 percent by volume and almost 40 percent by value of the total U.S. landings. This region encompasses the vast oceanic and coastal areas in the Gulf of Alaska and the Bering Sea, some of the world's richest fishing grounds. The fishing industry is Alaska's largest private employer, providing nearly 70,000 seasonal jobs, about 50,000 of them in the harvesting sector

Crew preparing longline for black cod fishery. The groundline, consisting of approximately 12,000 to 15,000 hooks spaced at intervals, is stored in tubs. Each tub may hold as much as 1,200 feet of groundline and 140 hooks. As many as 100 tubs of gear may be on board. The hooks are baited either while underway or, in the case of a short opening as is common in the North Pacific halibut fishery, prior to leaving port. (Herb Goblirsch, *Oregon Sea Grant*)

(Munro, 1990; McDowell Group, 1989; McDowell et al., 1989; Coughenower, 1987).

Vessel license data maintained by the Commercial Fishing Entry Commission in Juneau, Alaska, indicate that in July 1990, there were about 17,000 vessels licensed to fish commercially in four categories—fishing vessel, freezer/canner, tender/packer, and charter vessel. These numbers include vessels operating from West Coast ports and can vary significantly from year to year. Charter vessels, of which there are about 1,400, have expanded significantly in number over the past several years and are outside the scope of this study (some may be converted fishing vessels, but this cannot be determined from available data) (see Coughenower, 1986). There are about 175 freezer/canner and 1,400 tender/packer licenses; the remainder are fishing vessel licenses. The State of Alaska Fish and Game Processor Detail List for 1990, as of June 15, 1990, included about 200 vessels as combined catcher/processors (including those that process fresh fish) and 47 floating processors. Many of the tender/packers are fishing vessels employed full- or part-time to transport fish. About 80 percent of tender/packers are 78 feet or less in length; 37 percent are 40 feet or less.

Alaska's fishing industry can be divided into two parts—the nearshore, small-boat fleet and the larger, more highly capitalized vessels of the offshore fleet. The small-boat fleet concentrates on seasonal salmon, herring, halibut, black cod, and—to a lesser extent—Dungeness crab. In July 1990, there were 9,052 vessels licensed for the salmon fisheries, 2,162 in Bristol Bay alone. The size of vessels in some fisheries is strictly regulated by the state. Length limits for those in salmon fisheries are 32 feet in Bristol Bay and 58 feet in other areas. Vessels are typically owner-operated, 25 to 60 feet long, and valued under $500,000. Because overall length of vessels is limited under Alaskan conservation law, other dimensions are altered to accommodate greater carrying capacity. The groundfish fishery and most of the king and tanner crab harvests are conducted well offshore in the Bering Sea and Gulf of Alaska. Vessels in the groundfish fisheries are typically corporate-owned, 60 to 300 feet long, and valued between $1 million and $40 million (Coopers and Lybrand, 1990; Munro, 1990; McDowell Group, 1989; Natural Resource Consultants, 1988).

Vessels from ports in Alaska, Washington, and Oregon or from as far away as New England operate in the Gulf of Alaska and the Bering Sea. The fishing vessels used in these fisheries range from small, outboard-powered skiffs to large (300 feet or longer), sophisticated factory trawlers, crab vessels, and factory ships. In the latter examples, management is production and business oriented.

In this vast area, tenders, processing barges, power scows, and supply ships frequently operate in protected waters close to the fishing grounds to provide support for fishing activities. Many small ports receive varying volumes of products. Small, remote ports represent some of the most dangerous areas for fishing craft. High winds and waves at sea, sudden violent windstorms (williwaws or Taku winds) along the mountainous coastlines, rocky bottoms, winter icing conditions, and dense fog make operations in these waters dangerous and difficult. The nature of the region ensures that the captain and crew of all commercial vessels (big or small) must depend on onboard equipment and personal skills for survival. In these remote situations, surviving on shore can be as challenging as surviving at sea.

SUMMARY

Commercial fishing is an important, diverse, national industry with large numbers of vessels, high value produced, and many people employed. There are many kinds of fish, vessels, and fishing gear and techniques, but reliable data on the total numbers of vessels and fishermen are lacking. The operating environment is both dynamic and hostile. Economics, social organization, and human behavioral factors are complex, and external factors like fisheries management practices influence how, when, and where fishing is conducted. All these factors interact to form a mosaic in which safety is only one component. It is in this context that safety programs will need to be developed and implemented.

3

The Commercial Fishing Safety Record

This chapter analyzes the safety record of fishing industry vessels to better understand the nature, scope, and causes of safety problems and to provide a basis for assessing these problems in subsequent chapters. Safety problems include both vessel casualties (incidents involving actual or potential damage to fishing industry vessels) and also personnel casualties (fatalities and injuries). A basic issue is whether safety problems affecting fishing industry vessels and fishermen are significant enough to warrant government intervention beyond existing and planned safety regulations, voluntary efforts to improve safety, and provision of search and rescue (SAR) services.

DATA FOR ASSESSING FISHING INDUSTRY VESSEL SAFETY

Accurate historical and current data on vessels, fishermen, professional experience, hours and nature of exposure, and safety performance of personnel and equipment are fundamental to assessing safety problems, monitoring results of safety programs, and measuring the effectiveness of safety improvement strategies. Very few data are regularly collected or published on these parameters. The limited data make it difficult to quantify safety problems, determine causal relations, and assess safety improvement strategies. However, the data that are available indicate that significant safety problems exist and that human error, vessel and equipment inadequacies, and environmental conditions all contribute to them.

SAFETY DATA TERMINOLOGY

Vessel Casualties Incidents involving actual or potential damage such as capsizing, foundering, grounding, fires, and loss of propulsion or steering. Vessel casualties do not necessarily result in injury or loss of life.

Vessel Total Losses Vessel casualties resulting in the total loss of the vessel.

Fatalities Incidents involving loss of life.

Vessel-Casualty-Related Fatalities Fatalities resulting from vessel casualties, such as drowning resulting from capsizing.

Non-Vessel-Casualty-Related Fatalities Fatalities not resulting from vessel casualties, such as deaths resulting from falling overboard or getting caught in machinery.

Injuries Incidents resulting in nonfatal injuries. These may range from minor sprains and cuts to life-threatening injuries. Like fatalities, injuries may or may not be related to vessel casualties.

Personnel Casualties Fatalities or injuries.

Human Causes Causes of vessel or personnel casualties related to human factors.

Vessel Causes Causes of vessel or personnel casualties related to vessels or equipment.

Environmental Causes Causes of vessel or personnel casualties related to weather or sea conditions or to other factors affecting the operating environment, such as missing aids to navigation.

Fishing Fleet and Work Force Data

Detailed information about the composition amd utilization of the national fishing fleet and its work force is vital but not available. Such data are needed to reliably indicate the type and number of vessels and their relationship to vessel inspection, load lines, occupational safety and health regulations, and the way these factors relate to safety performance.

Fleet composition figures used in this report are estimates. Fleet data for the West Coast, Alaska, and Hawaii are better developed than for other regions. Close approximations of fishing vessels active in the West Coast region were determined for this study from landing records, referred to as "fish tickets" (Jacobson et al., 1990), and a fleet profile for Washington State fisheries (Natural Resource Consultants, 1988). For Alaskan fisheries, vessel permit data were provided by the Alaska Commercial Fisheries Entry Commission, Juneau,

AVAILABILITY OF FISHING SAFETY DATA

Vessel Casualties Coast Guard main casualty (CASMAIN) data are reasonably complete for major vessel casualties (those resulting in significant damage, vessel loss, or fatalities) for documented vessels. Data are incomplete for less-serious incidents on documented vessels. In addition, both major and less-serious vessel casualties may go unreported for undocumented (i.e., state-numbered) vessels. Coast Guard SAR data provide additional information on vessel casualties, but do not discriminate between documented and undocumented vessels.

Fatalities CASMAIN data are believed to be fairly complete for fishing vessel fatalities occurring at sea. However, some fatalities occurring close to shore or in inland waters may not be reported.

Personnel Injuries Data are insufficient to determine the total number of injuries that occur or injury rates, but limited insurance data provide information on injury patterns.

Fishing Vessels The number of vessels documented by the Coast Guard bearing fisheries license endorsements, although variable, can be closely estimated. Their physical condition and actual use in the fishing trade are not monitored. The actual number of vessels bearing state numbers that are engaged in commercial fishing is not known. Limited data are available regionally for vessel usage in commercial fishing and are generally more complete for vessels engaged in West Coast and Alaskan fisheries. Exposure and risk data are not available.

Fishermen The number of persons who engage full-time or part-time in commercial fishing is not known, with the exception of those in states such as Alaska where individual commercial or crew licenses are required. Demographic information is very limited. Fishing employment estimates are based on the estimated crew sizes of the estimated number of documented and undocumented vessels used in the fisheries trade. Exposure and risk data are not available.

available studies (Coopers and Lybrand, 1990; McDowell et al., 1989), and the committee's regional assessment (Munro, 1990). It appears that more complete data could be developed for Alaskan and North Pacific waters by analyzing and correlating vessel permit records and fish ticket data using automated data processing capabilities available through the Commercial Fisheries Entry Commission. Fish ticket and state license data on vessels categorized as fishing vessel, freezer/canner, or tender/packer could be correlated manually with federal documentation data in the Coast Guard's Marine Safety Information System (MSIS) and records of state vessel numbers (administered in Alaska by the

TABLE 3-1 Estimated National Fleet Size and Underway Work Force

Length (feet)	Number of Vessels	Positions per Vessel	Vessel Personnel Capacity
Documented Vessels	31,000		108,600
26-49	23,400	3	70,200
50-64	3,600	4	14,400
65-78	3,200	5	16,000
79+	800	10	8,000
Undocumented Vessels	80,000	1.5	120,000
All Vessels	111,000		228,600

Coast Guard). The level of effort required to do this was beyond the scope of this study. Well-developed employment estimates are available for Alaska fisheries (McDowell et al., 1989).

Table 3-1 presents the committee's estimates of the 1987 fleet size and a rough estimate of the work force capacity of the fishing fleet if all vessels were under way at one time. The total number of persons working as commercial fishermen in any given year is significantly higher (Chapter 2). The data presented in Table 3-1 are this report's benchmark for comparative analysis of safety data.

Crew sizes are very rough estimates based on assumptions. The actual number of persons aboard fishing industry vessels varies greatly. The actual crew size for vessels at least 79 feet long varies from about 4 to 5 for fishing vessels, to 20 to 30 for large catcher/processors, to well over 100 for the largest floating processor. There are no accepted industry averages. However, estimated crew sizes are available for Alaska (McDowell et al., 1989). For this analysis, the average crew size is assumed to be:

- for undocumented vessels, 1.5;
- for documented vessels, 3 for 26- to 49-foot vessels, 4 for 50- to 64-foot vessels, 5 for 65- to 78-foot vessels, and 10 for vessels of at least 79 feet.

Casualty Data

The best available data source about fishing industry vessel casualties is the U.S. Coast Guard's (USCG) main casualty (CASMAIN) data base. CASMAIN is well administered; however, the available data have various limitations discussed in this chapter and Appendix D. CASMAIN data are based on Coast Guard marine accident reports. Unless otherwise indicated, vessel casualty statistics presented in this report are derived from CASMAIN data for the years

PRINCIPAL U.S. COAST GUARD FISHING SAFETY DATA SOURCES

CASMAIN The Coast Guard's main casualty (CASMAIN) data base is based on Coast Guard Marine Accident Reports. By law, all incidents are supposed to be reported that result in loss or significant damage to vessels, loss of life, or serious injury. The data base provides information for vessel casualties and personnel casualties (fatalities and injuries). Relatively more information about the vessel, location of the incident, and environmental conditions at the time of the incident is provided for vessel casualties than for personnel casualties. In general, better data are available on safety problems for documented than for undocumented vessels.

SAR Data The Coast Guard's SAR data provide information on all SAR events in which Coast Guard forces provided some form of assistance, ranging from simple communications services to extensive surface and air searches. Although SAR data include less information than CASMAIN data, they provide useful additional information on the severity and location of SAR incidents involving commercial fishing vessels, and the general scope and nature of emergency events not meeting CASMAIN reporting thresholds.

SEER Data The Coast Guard's Summary Enforcement Event Report (SEER) data are the primary source for the scope of Coast Guard compliance examinations (boardings) of fishing vessels and boating safety violations reported for fishing vessels.

1982 to 1987. CASMAIN data are not presented for 1988 or 1989 because marine accident investigations, data coding, and data entry were still in progress during the study.

CASMAIN vessel casualty data are presented only for documented fishing industry vessels because marine accident reports are not available for an undetermined number of casualties involving state-numbered vessels. Data limitations precluded sorting of casualties by vessel employment—fishing, tender operations, or processing. Anecdotal information indicates that the majority of casualties are attributed to fishing operations. Some additional data, covering both documented and state-numbered vessels, are presented from the Coast Guard's SAR data base. CASMAIN, SAR, and other sources used in this study are described in the accompanying box and in Appendix D. Appendix E provides selected additional CASMAIN data tables. Developing reliable casualty rates and analyzing variations by region, fishery, gear type, or other criteria were not feasible with available data. A very rough estimate of casualty rates by vessel length for documented vessels is presented to provide a frame of reference for assessing the relationship of vessel size to safety performance.

CASMAIN data disclose that many, but not all, personnel casualties result from or are associated with vessel casualties. However, data on injuries are extremely limited from any source and are insufficient to develop even a rough estimate of personal injury rates aboard fishing vessels. The CASMAIN data base does provide complete enough information on fatalities to estimate fatality rates for both documented and undocumented vessels. Alaska Fishermen's Fund data, while a partial record of injury incidents, nevertheless provide a resource for general characterization of injuries in Alaskan commercial fisheries.

Most fishing industry fatalities are not reflected in published occupational injury and illness reports. Because most fishermen are self-employed, are not covered by workmen's compensation, and work aboard vessels with fewer than 11 employees, they are not included in the data collected and published for other industries by the Department of Labor and other federal organizations. The published national data on employment and occupational injury and illness fatalities aggregates fishing industry mortality statistics and rates with agriculture and forestry and covers only employers with 11 or more employees (see Appendixes D and G). National Safety Council (NSC) data also aggregate fishing with agriculture and forestry. These data, based on death certificates, are more complete nationally, but limit estimates to broad aggregations. Thus, occupational data for most personnel casualties attributable to commercial fishing are excluded from the principal sources of data available for comparing fatalities and injuries among industries. Using proportional mortality rates (PMR) for trend analysis and comparing mortalities in fishing with other industries is an alternative, but is currently available only for Washington State residents. PMR is discussed later in this chapter.

VESSEL CASUALTIES

Table 3-2 summarizes casualties to documented fishing industry vessels for the years 1982 to 1987. Four different measures are used:

- Number of vessel casualties. There were 6,558 vessel casualties reported, an average of nearly 1,100 annually.
- Number of vessel total losses. Of the 6,558 casualties to documented vessels, 1,298 (about 20 percent) resulted in total loss of the vessel. On average, more than 200 documented fishing industry vessels were lost each year. Vessel total losses were significantly lower in 1986 and 1987, but the information is insufficient to establish a downward trend.
- Vessel-casualty-related fatalities. A total of 348 fatalities (more than 50 per year average) resulted from casualties to documented fishing industry

TABLE 3-2 Casualties Involving Documented Fishing Industry Vessels: 1982-1987

Year	Number of Vessel Casualties	Number of Vessel Total Losses	Number of Fatalities Associated with Vessel Casualties	Total Vessel Damages (millions of dollars)
1982	983	259	33	93.6
1983	1,203	247	88	68.3
1984	974	241	52	71.0
1985	1,209	241	61	53.3
1986	1,095	151	46	39.5
1987	1,094	159	68	51.8
Total	6,558	1,298	348	377.5
Average	1,093	216	58	62.9

Source: USCG 1982-1987 CASMAIN data.

vessels; another 90 resulted from state-numbered vessel casualties; and 210 fatalities were not related to vessel casualty.

- Vessel damage. Total damage to documented vessels was $378 million, a loss of over $60 million annually. Total estimated damages were significantly lower during 1985 to 1987, but do not necessarily indicate a downward trend.

In assessing safety problems affecting fishing industry vessels, it is important to distinguish between these four measures. Each is important and can provide a different perspective on the nature and causes of vessel casualties. The number of vessel casualties is a measure of how many incidents occurred, regardless of how serious the consequences. The number of vessel total losses is a measure of how many incidents occurred in which the consequence—loss of the vessel—was very serious. Total vessel damages and the number of vessel-casualty-related fatalities measure the dollar and human costs. The total cost of human suffering associated with vessel casualties is considerably higher when personal injuries are considered, but the full scope of this aspect of the safety record cannot be ascertained from the data. Figures for total cost of vessel damages vary greatly, depending on which vessels were involved, the nature of casualties, and how damage estimates were derived. This limits the utility of damage estimates for comparative and trend analyses.

Nature of Vessel Casualties

Figure 3-1 depicts the nature of vessel casualties for each of the three vessel-related measures—number of casualties, vessel total losses, and fatalities. The data illustrate how greatly the picture of vessel casualties varies, depending on which measure is considered. For example, material failure accounted for

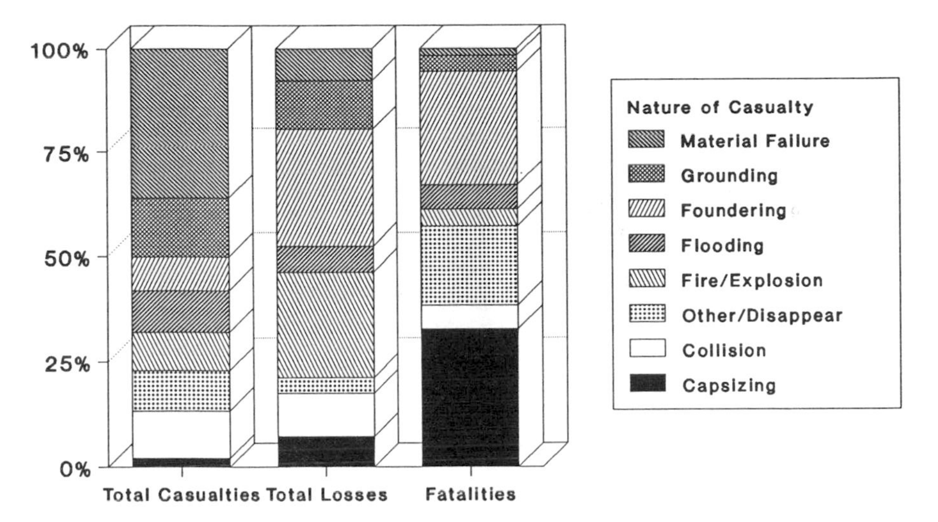

FIGURE 3-1 Nature of vessel-related casualties. Source: USCG 1982-1987 CASMAIN data for documented vessels.

36 percent of all vessel casualties, but only 8 percent of vessel total losses and 2 percent of vessel-related fatalities. Foundering or sinking accounted for only 8 percent of all vessel casualties considered together, but 28 percent of vessel total losses and vessel-related fatalities. Capsizing accounted for only 2 percent of vessel casualties and 8 percent of vessel total losses, but 33 percent of vessel-related fatalities.

These data demonstrate that no single type of incident stands out as the major safety problem in the commercial fishing industry. Instead, there are differing problems and consequences. Incidents like capsizing occur with less frequency, but have disastrous consequences. Others, such as material failure, may occur more frequently, but the consequences may not be as great.

Regional Distribution of Vessel Casualties

CASMAIN data disclosed significant regional variations in vessel casualties. (Except where indicated, casualty data for the Great Lakes and Hawaii are aggregated as "other" in regional distributions.) The West Coast region accounted for the greatest share of vessel casualties, vessel total losses, and fatalities (Figure 3-2). The North Atlantic region had the second highest share of vessel casualties, but ranked behind Alaska in vessel total losses and fatalities. There are also significant regional variations in the nature of casualties. For example, collisions account for over 18 percent of vessel total losses in the Gulf of Mexico compared with about 3.4 percent for Alaska, where grounding stood

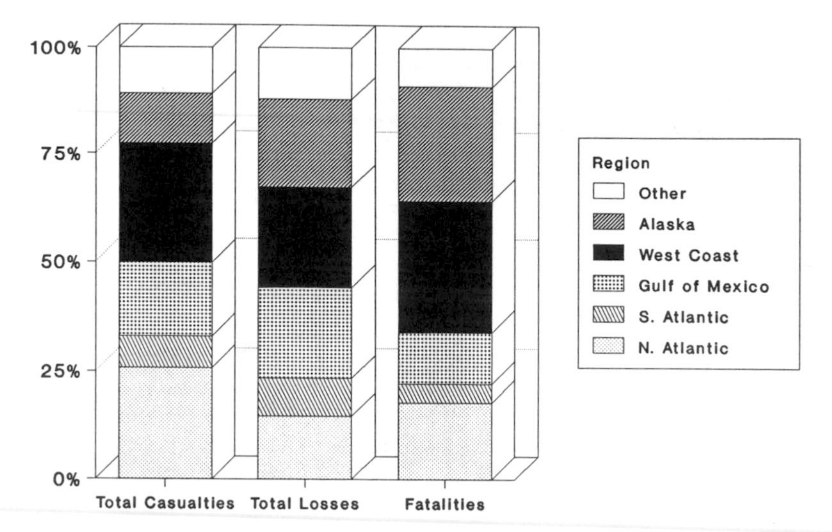

FIGURE 3-2 Regional distribution of vessel-related casualties. Source: USCG 1982-1987 CASMAIN data for documented vessels.

out in the data as a major factor. Whatever measure is used, it is clear that vessel casualty problems are not limited to one or two regions. Substantial vessel casualties, vessel total losses, and vessel-casualty-related fatalities occurred in all regions.

The consequences of vessel casualties also varied significantly by region. For example, vessel casualties in Alaska were three times as likely to lead to vessel total losses as in the North Atlantic, and resulted in three times as many vessel-related fatalities. However, it is difficult to tell to what extent this may simply reflect a higher reporting rate, or differences in fishing effort, gear, operating environment, and exposure to risk. Significant regional variations in the nature, scope, and consequences of vessel casualties indicate that reasons for these variations need to be identified and seriously considered during development and implementation of safety-improvement strategies and alternatives.

Casualty Distribution by Vessel Length

Vessel casualties occurred on vessels of all size classes (Figure 3-3). Vessels between 26 and 49 feet accounted for 47 percent of all casualties to documented vessels. However, because larger vessels carry more crew and represent a greater dollar investment, the consequences of casualties tend to be greater. Vessels 79 feet or greater in length accounted for only 12 percent of all casualties and 11 percent of vessel total losses, but they accounted for 24 percent of vessel-related fatalities and 47 percent of vessel damage.

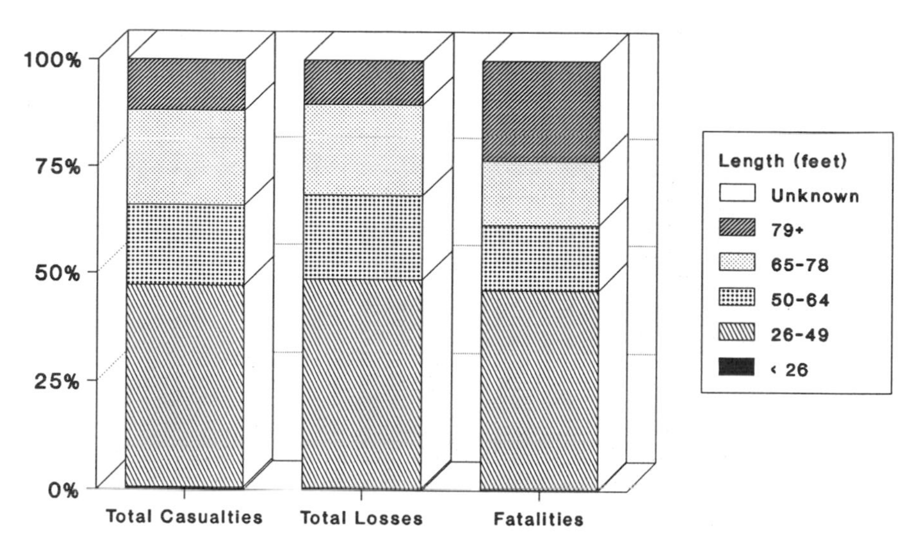

FIGURE 3-3 Distribution of vessel-related casualties by length. Source: USCG 1982-1987 CASMAIN data for documented vessels.

Casualty Distribution by Vessel Type and Usage

Casualties may vary by vessel type and usage. This cannot be directly ascertained from CASMAIN and MSIS data because of the absence of the necessary data fields. Although a resource-intensive process, some insight on this dimension could be developed by accessing CASMAIN data and casualty reports where names of vessels of certain types or engaged in certain activities, such as fish processing, are known. For example, a 1983 Coast Guard study used this methodology to determine that, from 1972 through 1982, the frequency of serious fires aboard fish processing vessels was dramatically higher than for other types of fishing industry vessels. During the period, fires led to 77 percent of processor losses but only about 19 percent of all documented vessel losses (USCG, 1983). It is not clear from CASMAIN data whether this anomaly has continued or, if so, to what degree. However, significant variations of this type are indicators that point toward problem areas meriting further attention.

Causes of Vessel Casualties

The CASMAIN data base includes data on the primary (proximate or most important) cause as well as secondary causes of vessel casualties. Appendix E (Table E-2) provides data on frequency for more than 80 primary causes. Causes were grouped into four broad groups for this report: human, vessel, environmental, and other or unknown causes. Human causes included general causes, such as "operator error," as well as a variety of specific errors, such

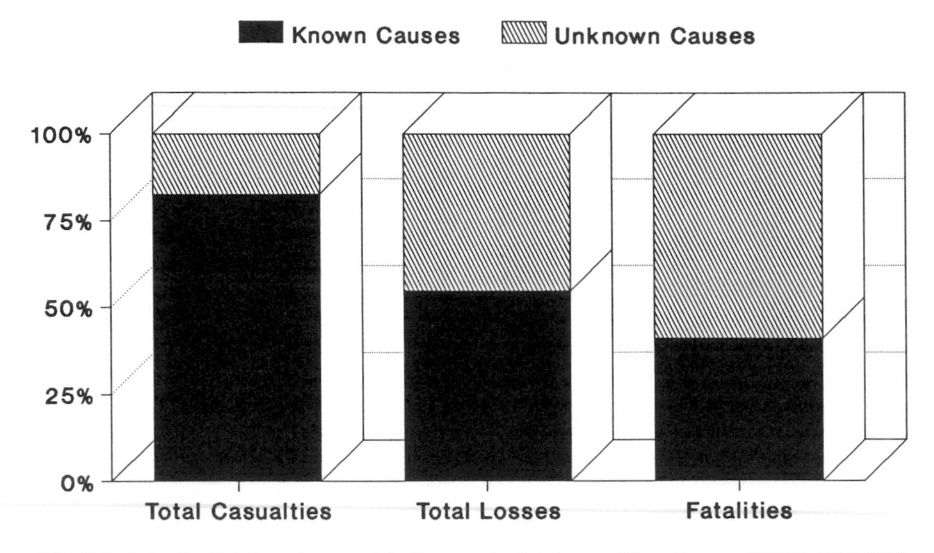

FIGURE 3-4 Availability of cause data for vessel-related casualties. Source: USCG 1982-1987 CASMAIN data for documented vessels.

as "failed to keep proper lookout." Vessel causes included various vessel or equipment problems, such as "failed structural materials" and "propulsion failure." Environmental causes included both "adverse weather or current" and others, like "submerged object."

There are two major limitations to what can be learned from the cause data in the CASMAIN data base. First, the cause data have large gaps in what is known (Figure 3-4). There is a high percentage of "unknown" as the recorded primary cause, especially for vessel total losses and vessel-casualty-related fatalities. This limits the utility of the data, since there is an insufficient basis to establish that the observed distribution of known causes would apply to incidents with unknown causes. Second, the cause data necessarily represent simplifications of circumstances leading to vessel casualties, based on subjective evaluation of complicated chains of events for which few data are available. Because of these limitations, the CASMAIN data provide only rough, but useful, indications of the relative contribution of different factors to safety problems.

Figure 3-5 shows that for vessel casualties with known causes, human causes played a role in 40 percent of casualties, 45 percent of total losses, and 61 percent of those related to fatalities. Similarly, where causes were known, vessel causes played a role in more than 62 percent of vessel casualties, 48 percent of total losses, and 40 percent of vessel-casualty-related fatalities.

As would be expected, the roles of human and vessel causes differed among incidents (Figure 3-6). Similarly, human causes were the primary cause of 68 percent of the vessel total losses resulting from grounding, but only 6 percent

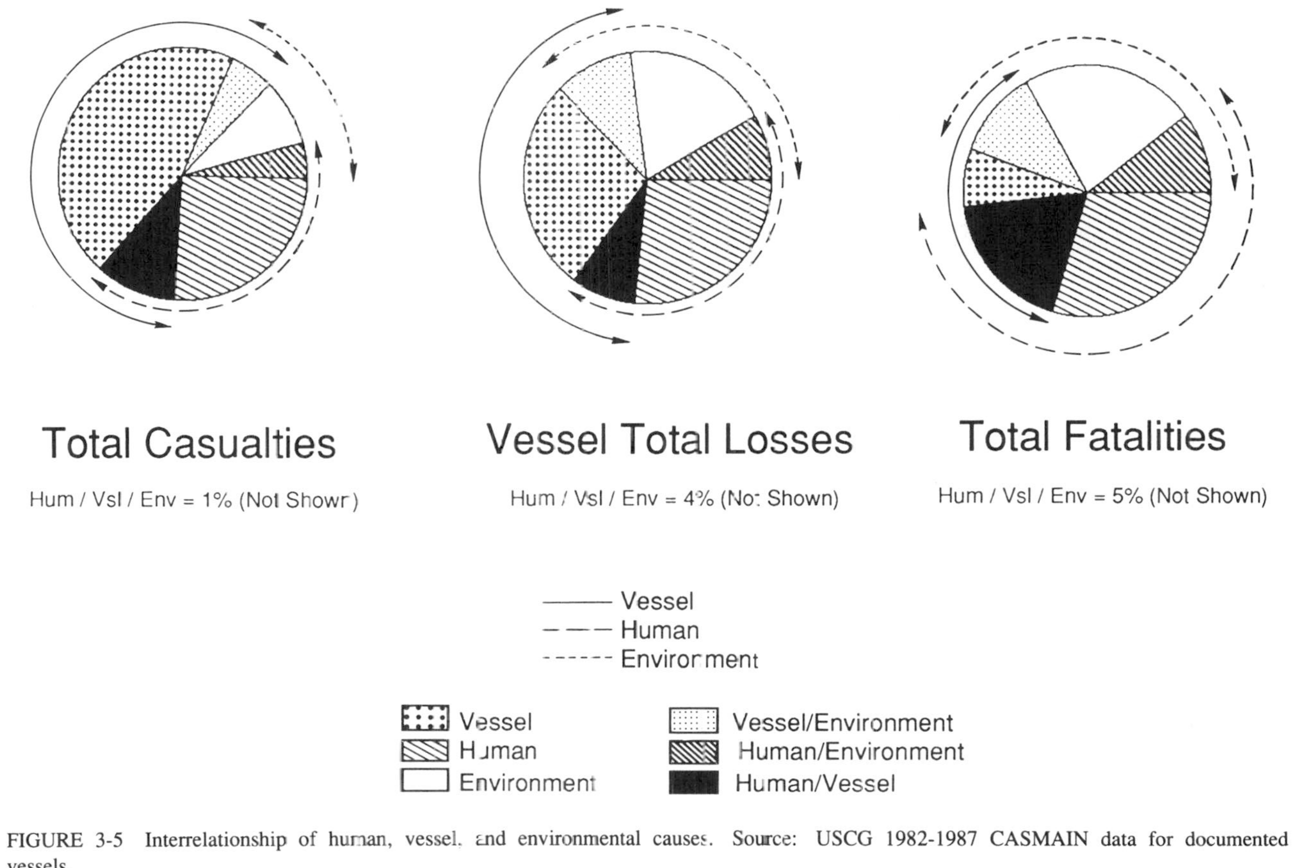

FIGURE 3-5 Interrelationship of human, vessel, and environmental causes. Source: USCG 1982-1987 CASMAIN data for documented vessels.

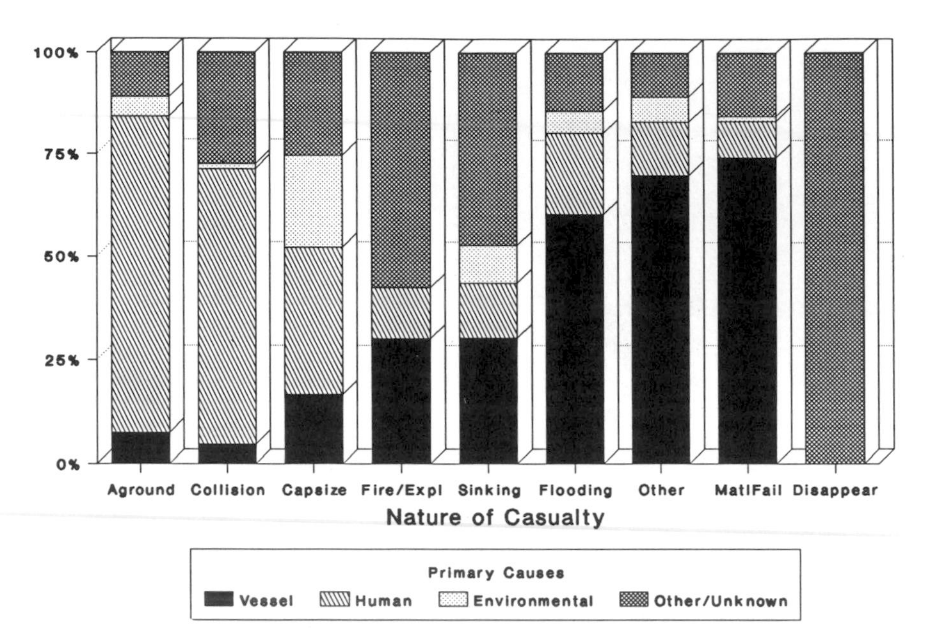

FIGURE 3-6 Primary nature and causes of vessel casualties. Source: USCG 1982-1987 CASMAIN data for documented vessels.

of vessel total losses resulting from fires and explosions. Vessel causes were the primary cause of 33 percent of vessel total losses resulting from flooding, compared with only 3 percent from collisions.

Vessel causes were the primary cause of about three-quarters of casualties involving material failure (Figure 3-7). From Appendix E (Table E-2), it appears that over 85 percent of known vessel-related causes can be attributed to structural, material, or system failures. This high percentage and the number of casualties of these types are reasons to closely examine safety-improvement alternatives that address the material condition of vessels and equipment.

Environmental causes were a factor in 20 percent of vessel casualties, 40 percent of total losses, and 47 percent of vessel-casualty-related fatalities where the causes were known. It is often suggested that commercial fishing is dangerous because of the extreme environment of the sea. Indeed, 17 percent of all casualties and 27 percent of all vessel total losses occurred when winds were higher than 20 knots. However, 34 percent of all incidents and 30 percent of vessel total losses occurred when wind strength was less than 10 knots. Thus, while environmental conditions were a factor in many casualties, they were just as clearly not a factor in even more.

Despite the limitations of cause data maintained in CASMAIN, they do support important findings. They illustrate that human, vessel, and environmental causes all contribute significantly to safety problems. This implies that there is a need to address all of these areas in searching for a complete solution. The data also show that causes of safety problems vary widely by region; they may also vary significantly by gear type, vessel configuration, or usage, as suggested by the high incidence of fires aboard fish processing vessels reported in the 1983 Coast Guard study. Factors unique to fish processing operations were implicated, including the routine presence of large quantities of combustible packing materials and hazardous chemicals, such as ammonia, and widespread use of highly flammable polyurethane foam insulation in refrigeration spaces. The study also implicated inadequate fire-fighting systems, inadequate safety procedures (e.g., no fire watches during welding and cutting), and little to no fire-fighting training of crews (USCG, 1983). It is clear from available evidence that safety problems cannot be attributed to a few universal causes common to all vessels, fisheries, or regions. Thus, causal relations need to be better understood and accommodated during development of safety-improvement options.

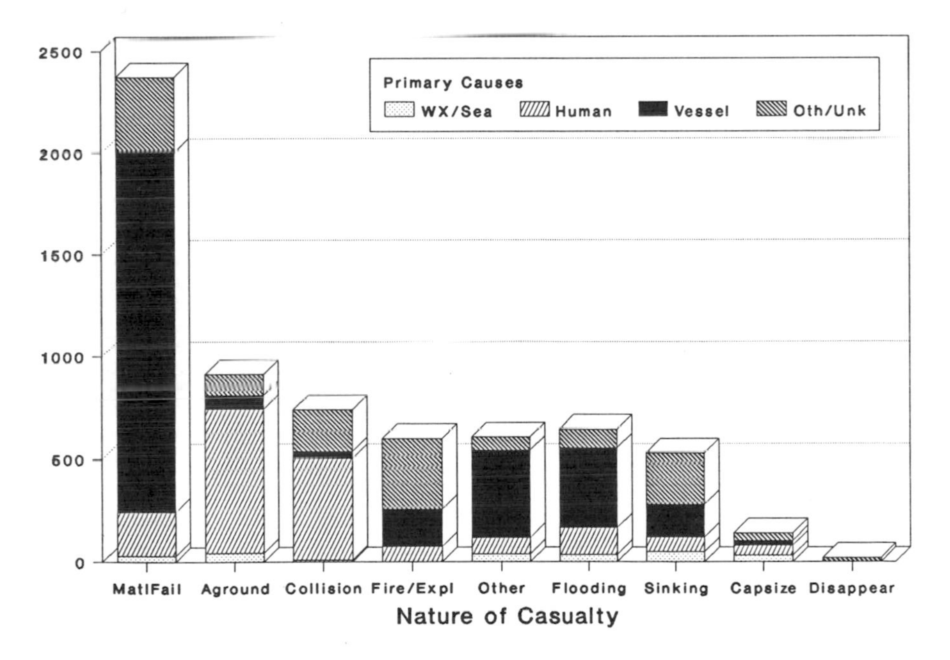

FIGURE 3-7 Number of vessel casualties by natures and primary causes. Source: USCG 1982-1987 CASMAIN data for documented vessels.

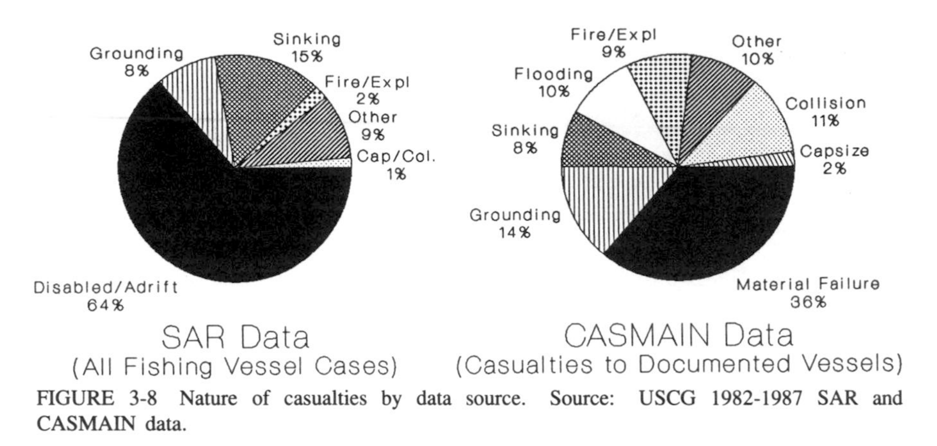

FIGURE 3-8 Nature of casualties by data source. Source: USCG 1982-1987 SAR and CASMAIN data.

Search and Rescue Data

Additional data on fishing vessel casualties are provided by the Coast Guard's SAR data base, but correlating them with CASMAIN is limited to general comparisons, for reasons discussed in Appendix D. Figure 3-8 compares the nature of vessel casualties reported in the CASMAIN and SAR data. "Disabled and adrift," roughly the SAR counterpart of CASMAIN's "material failure" category, represents an even higher share of all casualties in the SAR data—more than half. This finding must be balanced against the fact that SAR data contain a high number of minor breakdowns not threatening to life or vessels. Other casualty natures recorded in SAR data occur in proportions generally similar to CASMAIN data, except that relatively fewer collisions are recorded. Coast Guard assistance is often not requested in small-scale involvements of this type.

SAR data, like CASMAIN's, show differing cause patterns for different types of events (Figure 3-9). SAR data attribute 82 percent of disabled and adrift events and 72 percent of flooding and sinking events to vessel causes, including failure of the hull, propulsion systems, machinery, or other equipment. Incidences of nets or lines caught in propellers or rudders cannot be distinguished in the data. SAR data attribute 75 percent of groundings and 69 percent of collisions to personnel error.

Vessel Casualty Rates

Fishermen commonly consider inshore operations on small and midsized fishing vessels significantly less dangerous than offshore operations on larger vessels. SAR and CASMAIN data both reveal that the greater number of severe incidents involve vessels less than 79 feet long. Nevertheless, CASMAIN data suggest that larger vessels may be disproportionately represented in the casualty data in terms of rates of incidence. SAR data show a nominal increase in the percentage of more-severe incidents for vessels 40 feet or longer, but also indicate that life- and vessel-threatening events occur on all fishing grounds, with the heaviest concentrations inshore and on inland waters. Reliable casualty rates that accommodate exposure variables are needed to fully characterize the nature, scope, and relevance of these findings.

Casualty rates could serve as measures or indicators of safety performance, the magnitude of the problem, and causal relations among various factors affecting safety. With reliable rates, contributing factors could more easily be identified—nature of fishing activity, area fished, and vessel size. This could prove useful in directing safety-improvement measures where they are most needed. However, developing casualty rates for fishing industry vessels was severely hampered by the absence of reliable data on the number of fishing vessels, vessel material condition, fishing gear configurations, exposure variables, and other factors. Furthermore, there are far fewer larger vessels. Thus, one or two major casualties in any given year can significantly skew data and casualty rates. Nevertheless, it was possible to develop crude rates for documented fishing industry vessels by vessel length. It was also possible to identify from SAR data where the more severe incidents have traditionally occurred.

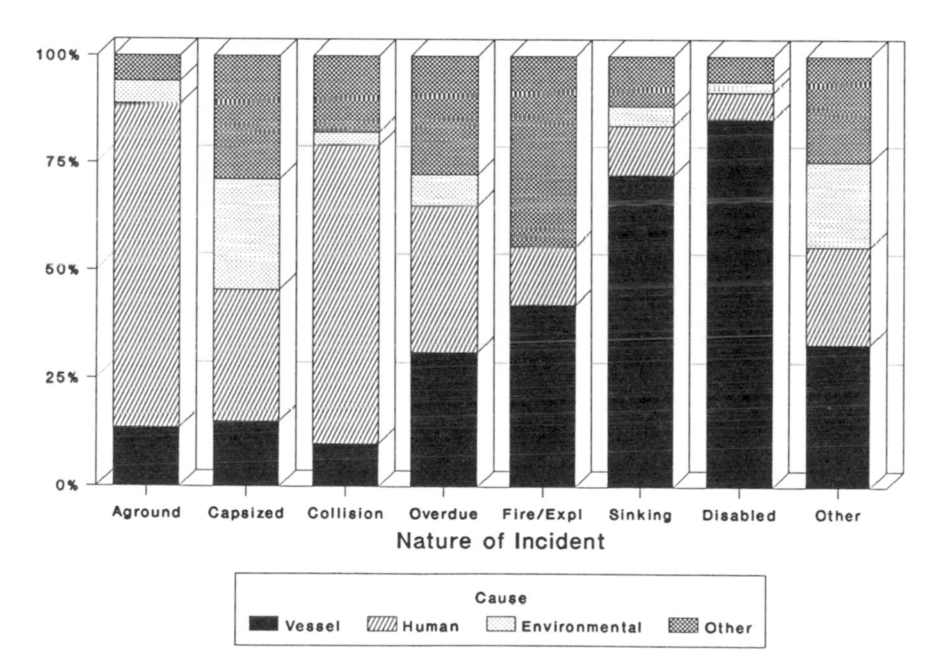

FIGURE 3-9 Distribution of causes of fishing-industry-vessel SAR cases. Source: USCG 1982-1987 SAR data.

TABLE 3-3 Estimated Vessel Casualty and Vessel Total Loss Rates (1982-1987)

Length (feet)[1]	Average Annual Vessel Casualties	Average Annual Vessel Total Losses	Number of Documented Vessels	Casualties per 1,000 Vessels	Total Losses per 1,000 Vessels
26-49	508	104	23,400	22	4
50-64	203	42	3,600	56	12
65-78	243	46	3,200	76	14
79+	130	23	800	163	29
All Vessels	1,084	215	31,000	35	7

[1]Excludes vessel casualties for which length was not known.

Table 3-3 provides a general indication of a relationship between vessel casualties and vessel length. The fleet size used is the 1987 estimate described earlier in this chapter. The actual fleet size for 1982 to 1987 is unknown, precluding trend analysis. Most importantly, the tables do not reflect fishing effort. Thus, the data and rates presented are not a sufficient basis for a rigorous safety program. No rates are presented for undocumented vessels. Even though the consequences of an accident involving a small fishing vessel might be as severe as those for one involving a larger one, annual losses may not reach marine casualty reporting thresholds. Furthermore, the estimate of undocumented vessels is very rough. Therefore, there is an insufficient basis to estimate casualty rates for state-numbered vessels.

The data and rates suggest that each year, about 3.5 percent of fishing industry vessels were involved in vessel casualties meeting reporting thresholds for marine casualties. Each year, slightly under 1 percent of vessels were involved in incidents resulting in total loss. Although vessels at least 79 feet long accounted for less than 3 percent of all documented vessels with fishing endorsements, they accounted for about 12 percent of all vessel casualties and 11 percent of vessel total losses. The data suggest that about 16 percent of documented vessels at least 79 feet long with fisheries endorsements were involved in vessel casualties each year, and almost 3 percent were total losses. This is significantly higher than for vessels under 79 feet.

Distribution of Search and Rescue Cases

Eighty-four percent of all fishing vessel SAR cases involving Coast Guard assistance occurred within 20 nautical miles of the coast. Thirty-five percent of the total were within 3 nautical miles of the coast, and 25 percent were in inland waters (Figure 3-10). The percentage distribution of SAR cases involving vessels shown in Figure 3-11 corresponds with the expected distribution of

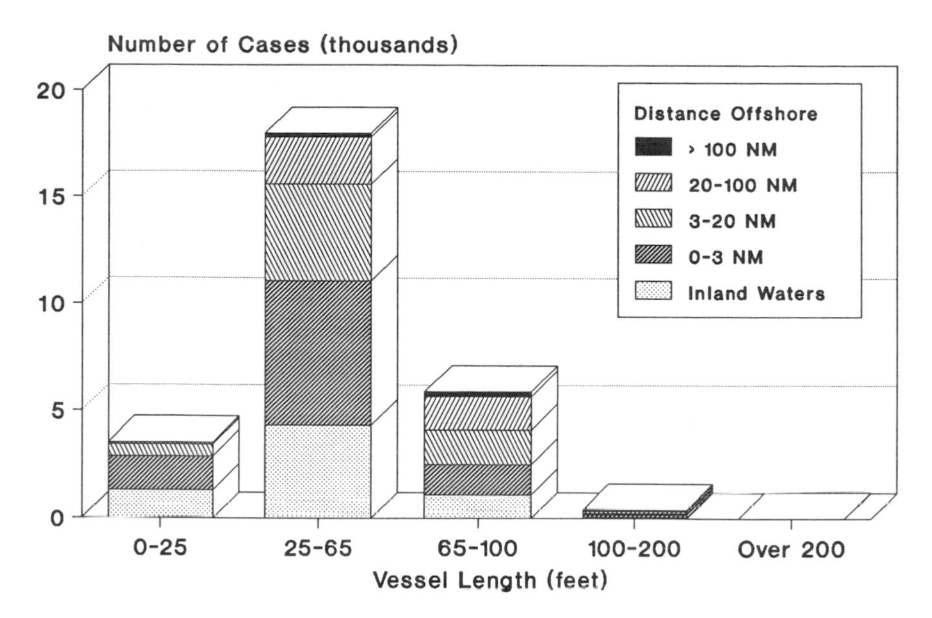

FIGURE 3-10 Distribution of fishing industry search and rescue cases. Source: USCG 1980-1987 SAR data.

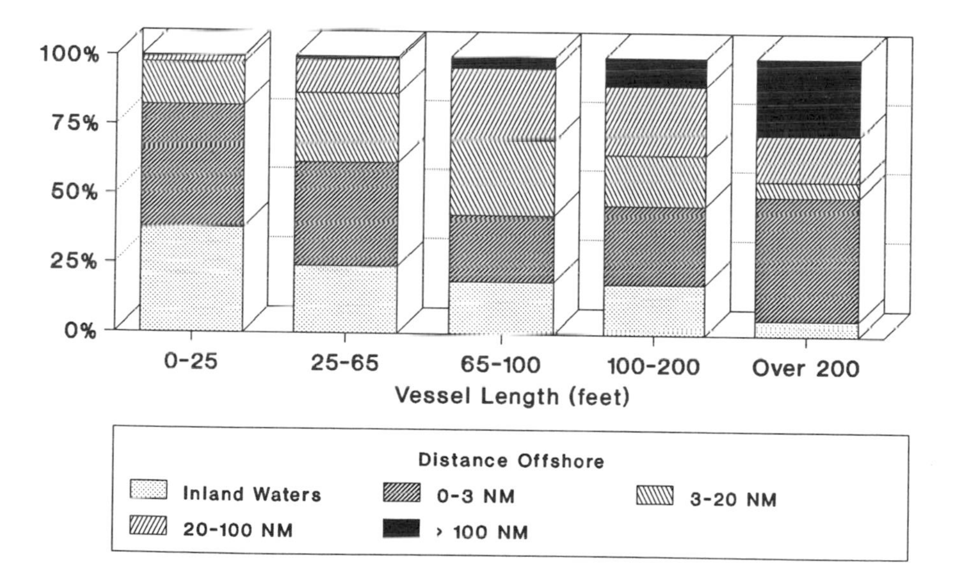

FIGURE 3-11 Distribution of fishing industry search and rescue cases. Source: USCG 1980-1987 SAR data.

fishing effort by vessel size in which the total number of involvements in offshore incidents increases by vessel size. The percentage distribution for vessels over 100 feet is based on a small number of incidents and is presented for completeness only.

Analysis of incident severity data revealed that 64 percent of cases in which lives or property were lost involved vessels between 26 and 40 feet in length. Vessels under 26 feet, especially those under 16 feet, were significantly less likely to experience a loss of life or property, while vessels 40 feet or longer experienced a small increase in the number of losses over that of smaller vessels.

As with distribution of SAR cases generally, density plots of SAR cases characterized as life- and vessel-threatening events show that they occur across the entire fishing grounds, and that a very large proportion of the total occur in sight of the shoreline. Figure 3-12 is a density plot for southern New England, depicting the number of incidents between 1980 and 1988 in which lives or property were lost or in danger of being lost. A similar density plot for the Pacific Northwest coast is shown in Figure 3-13.

Incidents involving personnel error were even more closely associated with operations close to shore. Where personnel or property were considered in danger of being lost or were lost and personnel error was listed as the primary cause, 92 percent of all incidents were within 20 nautical miles of the coast, 40 percent were within 3 nautical miles, and 38 percent were on inland waters. Although there are some regional variations, severe inshore and inland SAR cases tended to cluster around traditional fishing grounds, with heavy clustering found near inlets.

The concentration of fishing-vessel SAR events close to the coast can probably be explained by the fact that most fishing takes place relatively close to shore and that all fishing vessels must transit these waters. However, it is clear from the SAR data than even close to shore, fishermen are frequently involved in life- or vessel-threatening events.

Differences in Vessel Casualty Rates

Although the data are not conclusive, there appears to be some merit to the perception that large vessels operating offshore present more of a safety risk than smaller vessels operating inshore. The smaller population of larger vessels generally have longer periods of exposure than vessels operating inshore, many of which are operated on a part-time or casual basis. If larger vessels fish more intensively, even if casualty rates per days fished were the same for all vessels, larger vessels would be more likely to experience a casualty in any given year. Larger vessels fish farther offshore, in more severe environments. They may experience more normal wear, have more exposure to risk, and be farther away from help if an emergency develops. Finally, to an unknown extent,

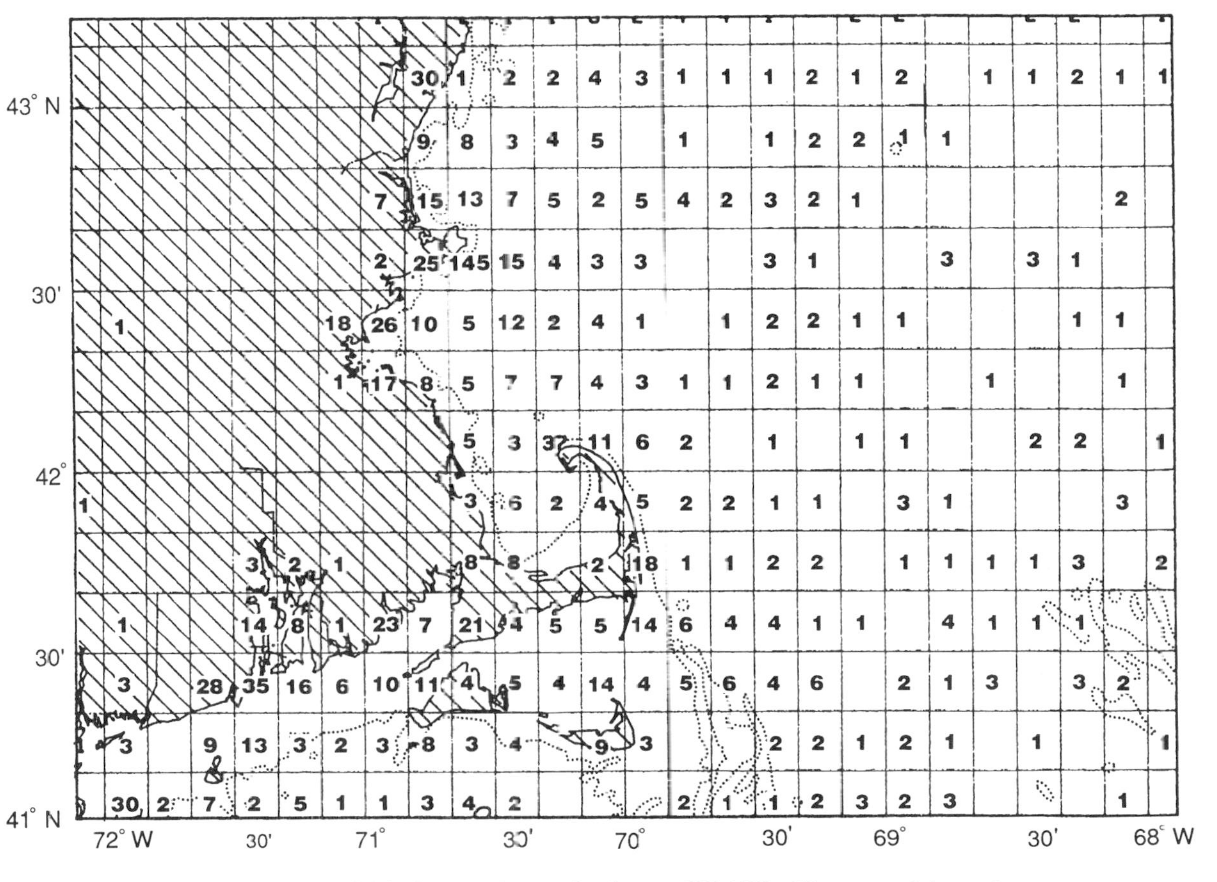

FIGURE 3-12 Commercial fishing vessel cases, fiscal years 1980-1988—life- or vessel-threatening events.

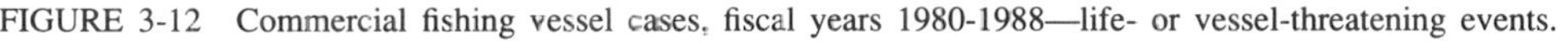

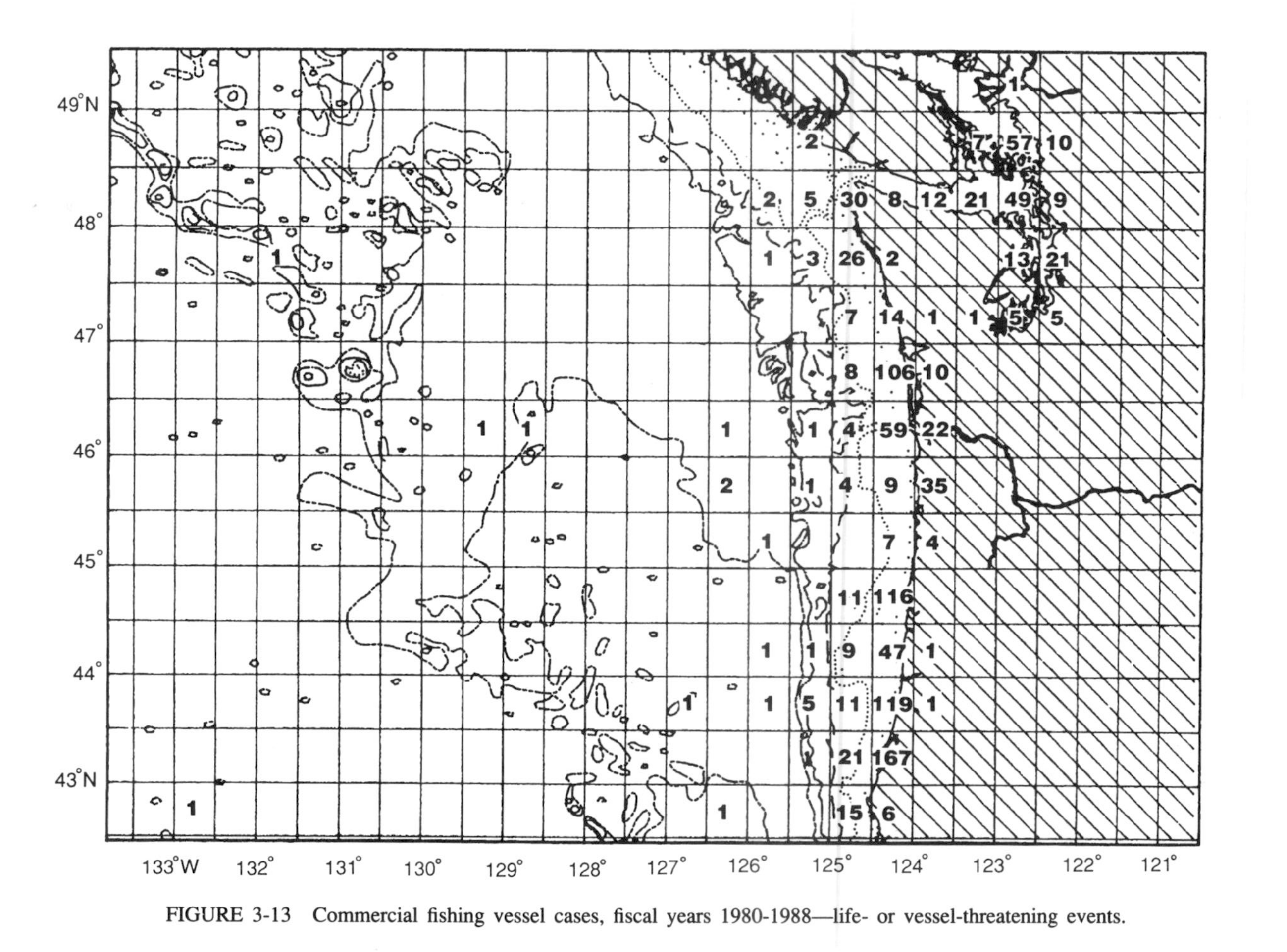

FIGURE 3-13 Commercial fishing vessel cases, fiscal years 1980-1988—life- or vessel-threatening events.

the difference in vessel casualty rates may reflect under-reporting of vessel casualties for smaller fishing vessels. What emerges is a picture of an industry in which the majority of events involve smaller vessels operating inshore or on inland waters but are distributed among a larger population of fishing vessels. The larger vessels reflect a relatively smaller number of involvements overall, yet are more likely to be involved in an incident.

FATALITIES

According to the CASMAIN data base, there were 648 commercial fishing fatalities from 1982 through 1987, an average of 108 per year (Table 3-4). Of these, the majority, 348, were associated with casualties to documented vessels. There were also a large number that were not related to vessel casualties, such as falls overboard. About 14 percent of fatalities were associated with state-numbered vessels. The number of fatalities was approximately evenly distributed between the Atlantic Coast, Gulf Coast, West Coast, and Alaska (Figure 3-14). More than 60 percent of all vessel-casualty-related fatalities resulted from capsizing and foundering (Figure 3-15). More than two-thirds of deaths not related to vessel casualties resulted from falling into the water or "vanishing" (Figure 3-16).

For more than half of all fatalities recorded in CASMAIN data, the primary cause was coded as "unknown" (Figure 3-4). "Human causes" were the primary cause in nearly two-thirds of the fatalities for which causes were coded as known, and "vessel causes" accounted for about one-fifth (Figure 3-17). Thus, as with vessel casualties, the limited available data suggest that

TABLE 3-4 Commercial Fishing Fatalities

		Vessel-Casualty-Related Fatalities		
Year	Total Fatalities	Associated with Documented Vessels	Associated with Undocumented Vessels	Fatalities Not Related to Vessel Casualties
1982	89	33	8	48
1983	145	88	27	30
1984	95	52	16	27
1985	105	61	16	28
1986	98	46	9	43
1987	116	68	14	34
Total	648	348	90	210
Average	108	58	15	35

Source: USCG 1982-1987 CASMAIN data.

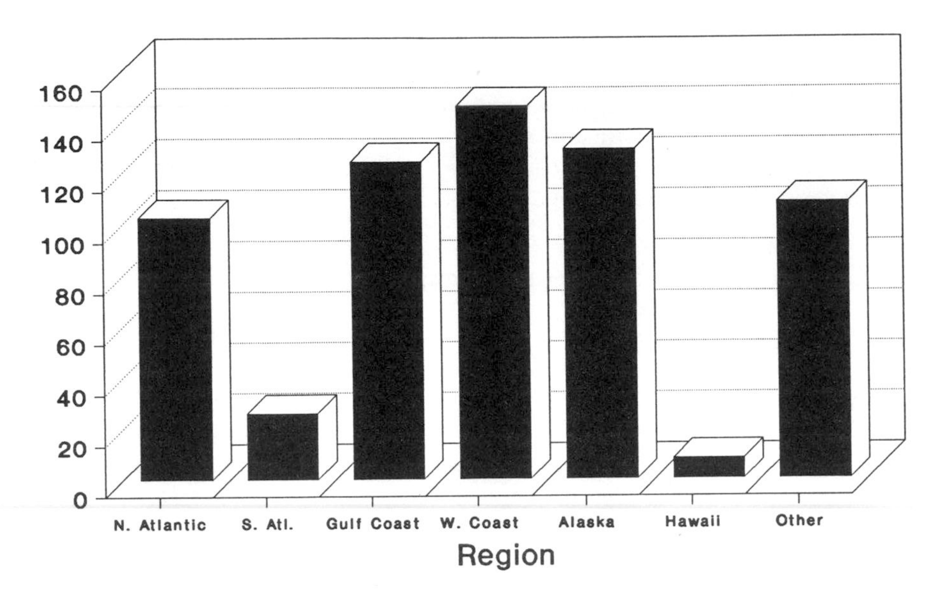

FIGURE 3-14 Commercial fishing fatalities: all fishing industry vessels. Source: USCG 1982-1987 CASMAIN data.

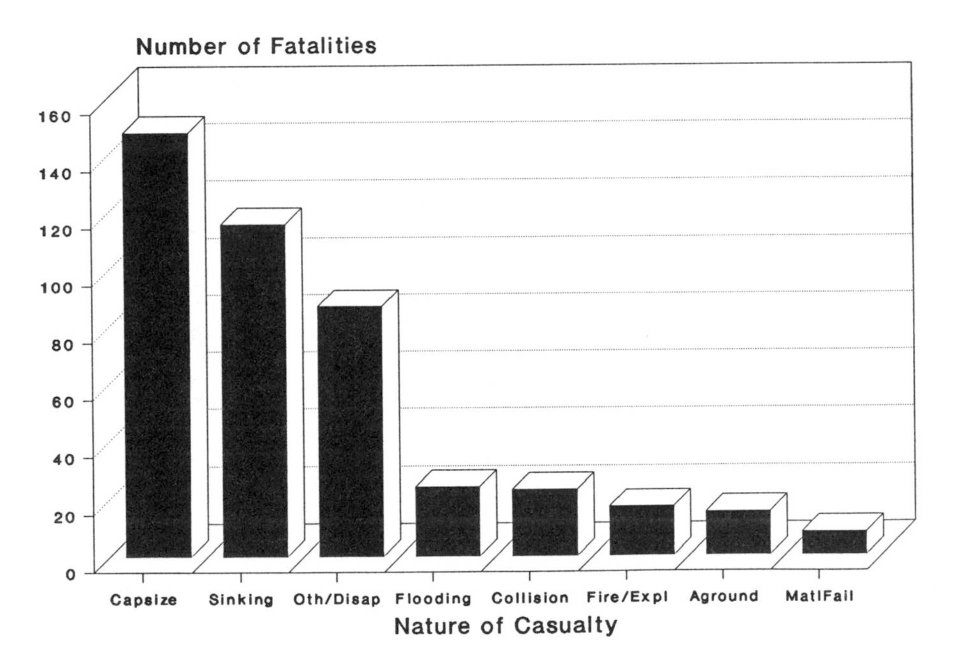

FIGURE 3-15 Vessel-related fatalities by nature of casualty. Source: USCG 1982-1987 CASMAIN data.

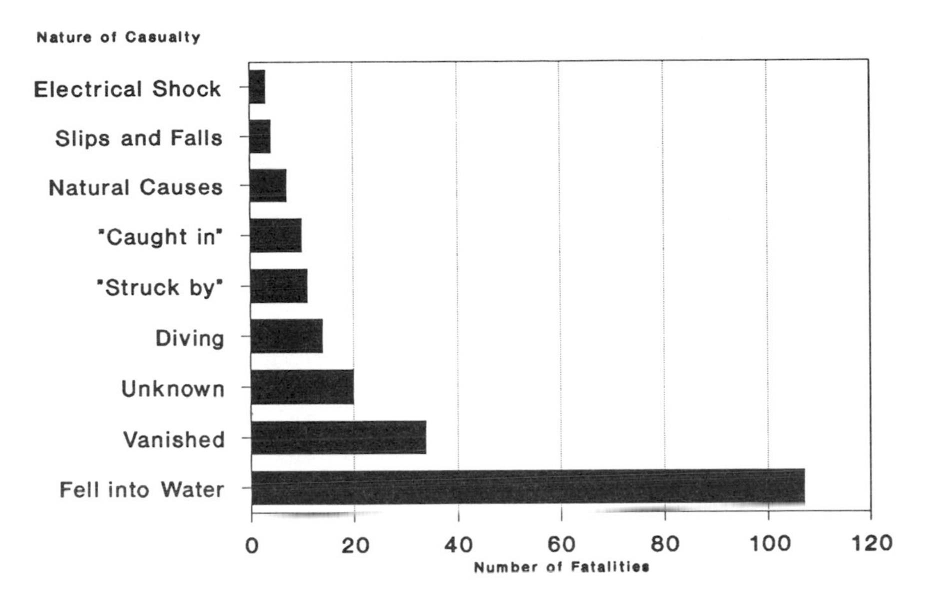

FIGURE 3-16 Non-vessel-related fatalities by nature of casualty. Source: USCG 1982-1987 CASMAIN data.

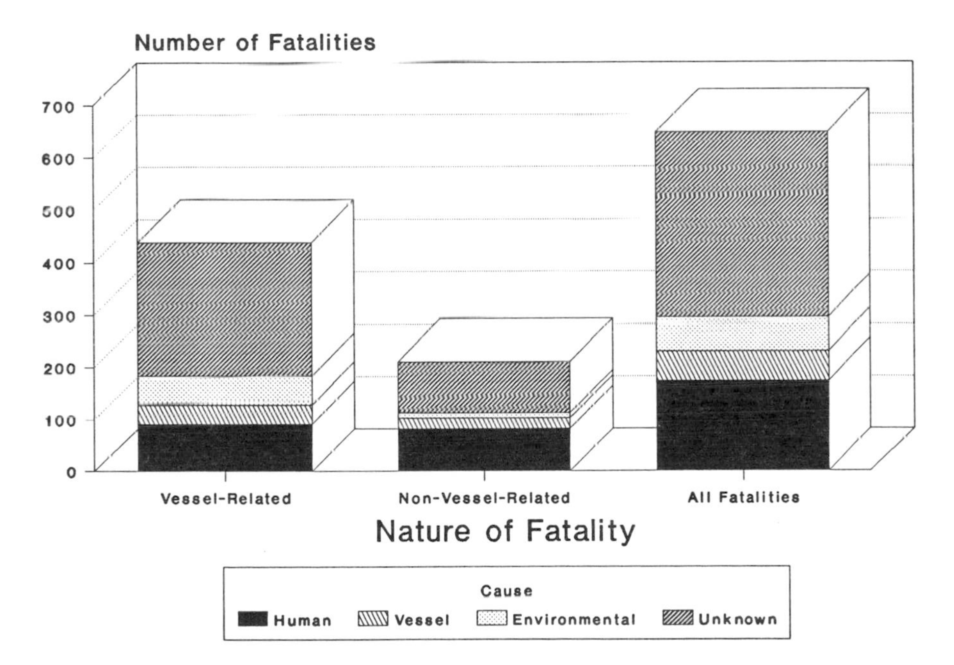

FIGURE 3-17 Primary causes of commercial fishing fatalities. Source: USCG 1982-1987 CASMAIN data.

strategies targeting human, vessel, equipment, and environmental factors may be needed to reduce mortality in the commercial fishing industry.

Fatality Rates

To compare the safety record of commercial fishing with that of other industries and to further examine the relationship between safety and vessel size, the committee estimated fatality rates. This effort was hampered by the absence of reliable data on commercial fishing employment for most fisheries. Estimates from Table 3-1 were used. Because of methodological problems, none of the fatality rates presented are precise estimates. Rates for fatalities associated with undocumented vessels are also presented, but are of limited utility because of uncertainty about actual numbers of vessels and fishermen.

The only known facts presented in the tables and figures are the number of fatalities recorded in CASMAIN. The fatality data do not discriminate among functional responsibilities; thus, they reflect all personnel aboard. Coast Guard officials believe that most commercial fishing fatalities are captured in CASMAIN data, principally because Coast Guard policy permits issuance of letters of presumptive death needed to settle estates when remains of fishermen lost at sea are not recovered. The reliability of CASMAIN fatality data could not be evaluated, except in Washington State, where state vital statistics were compared with CASMAIN data as part of the West Coast regional assessment. Nearly all commercial fishing fatalities for Washington State fishermen were reflected in CASMAIN data. Other problems in the analytical technique include the use of estimates of variable reliability, nonavailability of an average fleet size for the period, and the lack of exposure data, which prevents normalizing the data by fishing effort in different fisheries or by vessel or gear categories.

One perspective may be obtained by examining the distribution of fatal incidents (i.e., an incident in which at least one person died). While over 53 percent of fatal incidents (231 of 434) were associated with vessels under 50 feet long, fatal incident rates per 1,000 vessels grew substantially with increased vessel length (Figure 3-18). The data revealed that for vessels under 50 feet long, such incidents were more likely to result from a casualty to the vessel. The inverse was evident for vessels at least 65 feet long; fatal incidents not related to a vessel casualty were dramatically higher. The rates suggest that different safety problems resulting in fatal incidents exist for small and large vessels. For example, the higher rate for fatal incidents coincidental with casualties to smaller vessels may indicate a need for survival equipment not typically carried. The substantially higher frequency of non-vessel-casualty-related fatal incidents associated with larger vessels suggests that factors such as occupational safety issues merit attention.

Examining the estimated number of fatalities (from all causes) per fatal incident (Figure 3-19) gives another perspective. Since larger vessels have

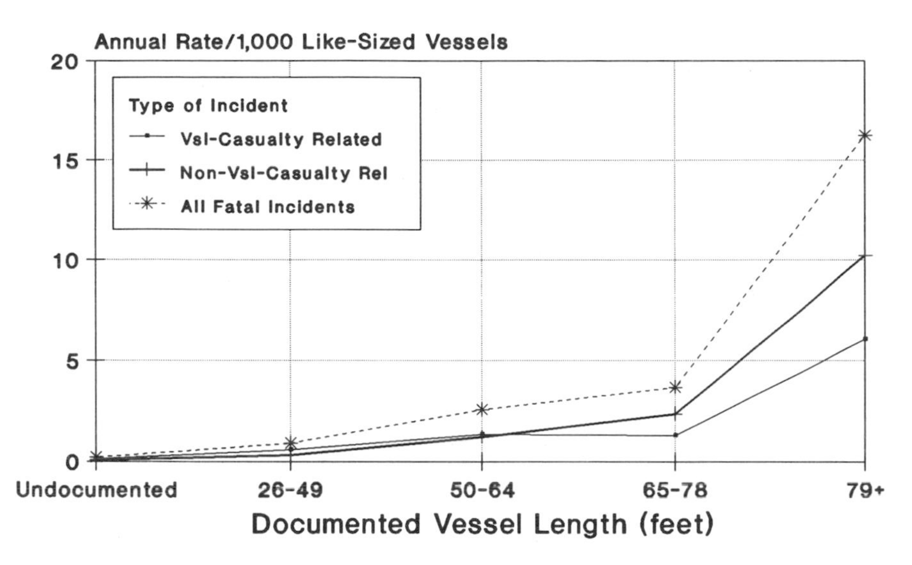

FIGURE 3-18 Commercial fishing fatal incident rates by vessel length. Source: USCG 1982-1987 CASMAIN data for all recorded fatal incidents.

higher fatality incident rates and larger crews, the number of fatalities per incident involving such vessels might be expected to be substantially higher. Indeed, when fatalities related and not related to vessel casualties are separated, the average fatality rate per incident increased modestly for vessels at least 65 feet long. The data for the 65- to 78-foot category are skewed by one incident in which there were 13 fatalities. Removing this one incident results in a 2.4 fatalities per incident rate for vessels in this length category. Although the trend is upward for larger vessels, it is not dramatic. Anecdotal information shows that in many cases, some of the crew are rescued, often because there were survival systems on board.

An even more striking picture relevant to vessel length emerges when the population of vessels and persons aboard fishing industry vessels is an element of the analysis (Table 3-5). The largest number of vessel-related fatalities, over 50 percent, involved fishing industry vessels under 50 feet. However, the relative number of incidents by vessel length is much greater for larger vessels, with correspondingly higher estimated fatality rates. The higher rates may be partly explained by more hours of exposure to risk than those for inshore boats, due to time in transit and time fished, difficulty in seeking shelter to escape rapidly deteriorating environmental conditions, complex fishing gear, and distance from Coast Guard rescue resources. These factors tend to contrast with the nature of inshore fishing. Even though the data are not available to normalize exposure rates, the marked jump for vessels at least 79 feet long strongly indicates that there is a greater likelihood that persons aboard these

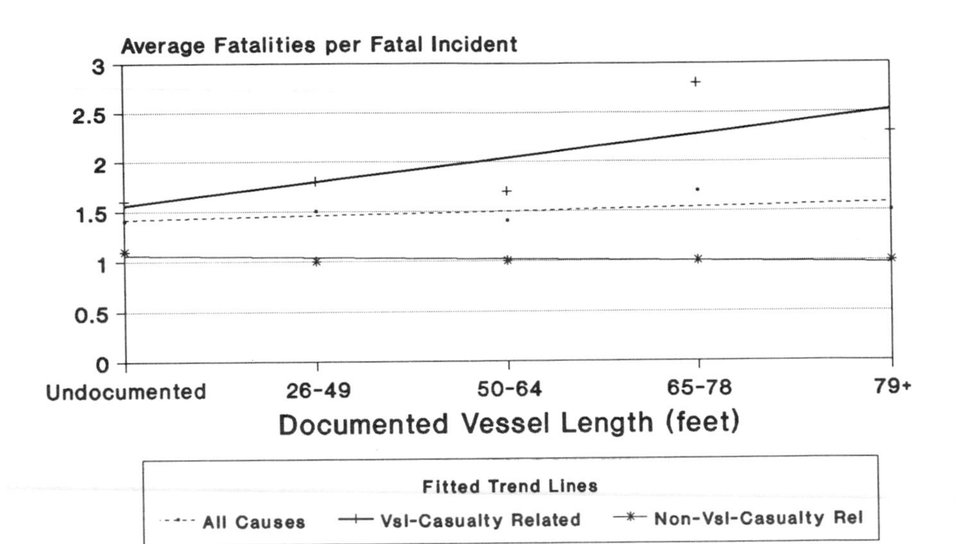

FIGURE 3-19 Distribution of fatalities per fatal incident. Source: USCG 1982-1987 CASMAIN data.

TABLE 3-5 Estimated Commercial Fishing Fatality Rates (All Fatalities)

Vessel Category	Total Fatalities	Annual Fatalities[1]	Employment (thousands)[1]	Annual Fatality Rate per 100,000 Workers[1]
All Vessels	648	108	229	47
Documented Vessels (feet)	507	85	109	78
26-49	198	33	70	47
50-64	76	13	14	88
65-78	116	19	16	121
79+	117	20	8	244
Undocumented Vessels	141	24	120	20

[1]Rounded numbers.

Source: USCG 1982-1987 CASMAIN data.

TABLE 3-6 Selected U.S. Industrial Fatality Rates (Estimated)

Industry	Annual Fatalities[1]	Employment (thousands)[1]	Annual Fatality Rate per 100,000 Workers[1]
Documented Fishing Industry Vessels	85	109	78
Undocumented Fishing Industry Vessels	24	120	20
All Fishing Industry Vessels	108	229	47
Mining	430	920	47
Construction	2,230	5,940	38
All industries	11,240	106,280	11

[1]Numbers estimated and rounded.

Source: USCG 1982-1987 CASMAIN data.

vessels will be involved in a life- or vessel-threatening event than will their counterparts on smaller fishing industry vessels.

Comparing fatality rates with those of other industries is severely limited by availability and limited value of data. For example, since many fishermen are on their vessels for extended periods of time when they are not actually engaged in fishing, they are often exposed to operational and occupational hazards for a longer time than workers in other industries are. Exposure data for such dimensions are not available. Nevertheless, the data that are available (Tables 3-5 and 3-6; Knapp and Ronan, 1990) suggest that fishermen aboard documented vessels perish at extraordinarily high rates and are thus more likely to die on the job than workers in most other industries are. Thus, concern for fishermen's welfare is well founded, considering safety performance within the industry. The aggregate fatality rate presented for U.S. commercial fishing (47 per 100,000 workers) is comparable to the mortality rate (45.8 per 100,000) for commercial fishermen in Canada's Atlantic provinces (Hasselback and Neutel, 1990).

Proportional Mortality Rates

A different approach for comparing fatality rates among industries, which does not require employment data, is the use of proportional mortality rates (PMR). PMR as a monitoring technique for mortality trends overcomes lack of reliable population-at-risk data by basing comparisons between occupations on the percentage of deaths resulting from occupational causes as a share of total deaths. Milham reports that there is very close correlation between PMR and standardized mortality rates (SMRs) and that this finding has been established through population-at-risk studies in other occupations within Washington State

(Samuel Milham, State of Washington Department of Health, personal communication, 1990).

PMR has been applied to mortality data recorded on death certificates of fishermen domiciled in Washington State (regardless of the location of death). Residents whose occupation was listed as fisherman had a high incidence of occupational deaths. PMR shows no significant change since the mid-1950s in accidental death trends among state commercial fishermen. Drowning was the predominant cause of accidental death listed on death certificates. Two-thirds of the time, the vessel associated with the death was intact. Only 34 percent of the time did it founder or capsize. Most alcohol-related deaths occurred while vessels were at the dock. However, this finding is biased, because a body must be available for analysis; this often is not the case for fatalities at sea. Milham reports that authorities in Great Britain also apply PMR to the fishing industry with similar findings.

PMR ranked occupational mortality of state residents in the following order:

- blasters and powdermen;
- loggers;
- log truck drivers;
- construction engineers;
- telephone and power linemen; and
- fishermen and oystermen.

Thus, for Washington State fishermen, PMR discloses that several other occupational fields are somewhat more dangerous, although not by a great deal.

From death certificate information, Milham also ascertained that the general locations of fatalities were as follows:

- in the ocean—30 percent;
- in sounds or bays—39 percent;
- in rivers—13 percent (largely associated with tribal fisheries); and
- in inlets—9 percent.

These percentages are consistent with the SAR data presented earlier, which show that a high number of life-threatening events occur inshore and on inland waters.

There are limitations to the usefulness of PMR for assessing commercial fishing risks. This methodology aggregates all fishermen as a common group, regardless of where or how often they fish. Occupation information from death certificates is not computer coded for many states, and they usually provide only limited information on circumstances. Furthermore, employment as a casual or part-time commercial fisherman may not be reflected in occupational data on the certificate. Nevertheless, PMR could provide a useful methodology for tracking mortality trends in commercial fishing in different coastal states. Although resource intensive, in-depth analysis and correlation of death certificate data with other safety information could lead to better insight on contributing causes to fatalities.

Relationship of Fatalities to Survival Equipment

CASMAIN contains "lack of available PFD" (personal flotation device) as a cause category for personnel casualties, interpreted by the Coast Guard to mean that a PFD was, for whatever reason, not available to the individual. CASMAIN data recorded only one fatality in which failure to use a PFD was recorded as the primary cause. However, anecdotal information frequently discloses that personnel lost overboard and not recovered were not wearing PFDs. The infrequent use of "lack of available PFD" for primary cause probably reflects the fact that this type of deficiency is more likely a contributing factor rather than a primary cause. Neither CASMAIN nor SAR data record what types of lifesaving devices were available, which if any were used, or how the devices performed during fatality or personnel injury incidents. This is a major data gap that forces all analysis of survival-equipment effectiveness to use less-inclusive information resources, such as accident investigations and anecdotal information.

PERSONAL INJURIES

As noted earlier, insufficient data are available on which to base even a rough estimate of how many injuries occur in the commercial fishing industry or to what degree injury rates may vary by vessel length, deck layout, or other factors. However, there are sufficient data to indicate the kinds of injuries and causes of greatest concern. For Alaskan commercial fisheries, Alaska Fishermen's Fund data can be used to identify possible variations by vessel and gear type, fishing district, nature of injury, body part injured, and Alaska residency status (Alaska Department of Labor, 1988).

The most definitive research relevant to injuries was accomplished by Nixon and Fairfield (1986), who examined a random sample of commercial fishing insurance claims. Slips and falls constituted the largest number of injury cases, more than 25 percent of all permanent injuries. Falls overboard and crushing injuries were also common. These injuries were typically caused by handling fishing gear on deck, most frequently on vessels equipped with heavy gear manipulated with powerful winches. The most common types of crushing incidents occurred as a result of a hand, leg, or body caught or wound in a winch; trawl doors, traps, or other heavy gear dropped on feet; and being crushed against a bulwark or other fixed object by heavy swinging gear. A classic scenario, repeated again and again, was of a crew member attempting to

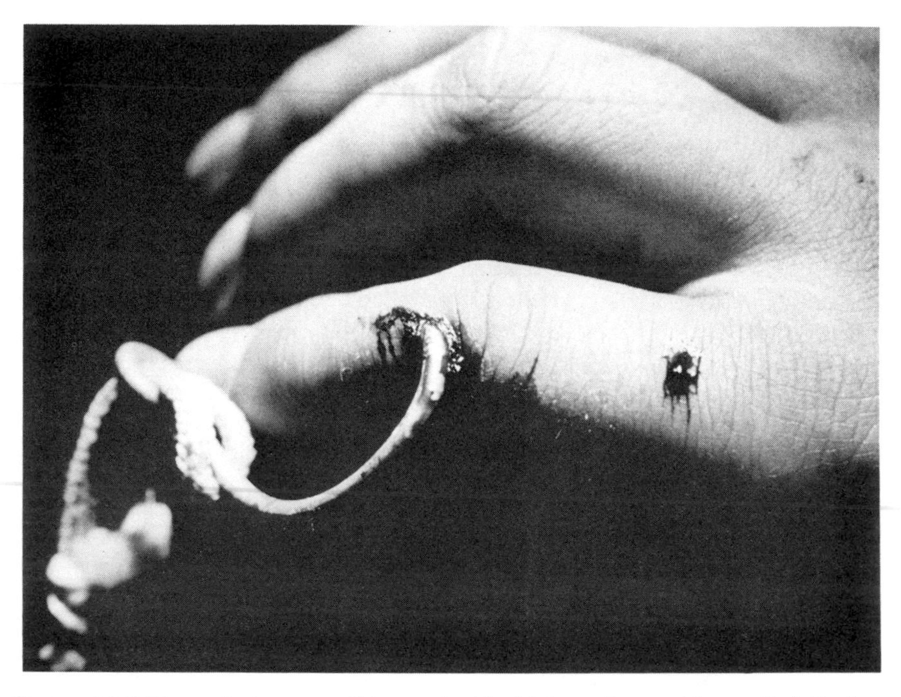

Commercial fishing routinely exposes fishermen to risk of injury, often requiring medical attention ashore. The scale of personal injury incidents in the fishing industry is not known but is thought to be large relative to injuries experienced in most other industries. (Art French, M.D., *Alaska Sea Grant*)

adjust a moving cable with hands or feet, losing balance on a slippery, pitching deck, and being wound into the winch before the power could be cut off.

An example of the possible extent of injuries is found in records of the Alaska Fishermen's Fund, under which Alaskan fishermen are eligible to receive up to $2,500 in compensation for fishing-related medical expenses. Although fund statistics are not routinely published, some statistical information is available. In fiscal year 1987 (FY 87), the fund recorded 2,363 personal injury claims (1,867 were approved). Roughly 1 in 20 fishermen eligible under the fund filed claims. Where gear type was listed in the claims, longline gear was the leader (34 percent), followed by seine (22 percent) and gill net (20 percent). The Fund uses a unique injury-classification system, which constrains comparison with injury data collected for other industries. For FY 87, 32 percent of claims filed were classified as sprains or strains and 27 percent as cuts, lacerations, abrasions, or punctures. Injuries involving fingers, hands, arms, or shoulders were most common (41 percent) and were most frequently associated with longline gear; cuts and infections were the leading "nature of injuries" in these cases. A comparison of Alaska Fishermen's Fund data with Alaska commercial fishing license data is not available. Rates of injury incidents by gear type are not known.

The Commercial Fishing Industry Vessel Safety Act of 1988 (CFIVSA) calls for insurance industry procedures to collect better information on injuries. However, even these data will be limited because of the nature and complexity of the international insurance market (Chapter 7) and the fact that some owners, particularly those on smaller vessels, operate without insurance.

HUMAN FACTORS DATA

CASMAIN is the most complete source of human factors data for documented U.S. vessels, including documented fishing vessels. The data have significant shortfalls, which are already being addressed through research and development by the Coast Guard. Nevertheless, CASMAIN data suggest that certain human attributes are more prevalent than others as primary and secondary causes of casualties. Human factors as causes of accidents are addressed in Chapter 5.

STRATEGIES FOR IMPROVING SAFETY DATA

Fully effective administration of safety programs depends on adequate data resources. Without reliable and statistically valid data, safety shortcomings cannot be identified with clarity, and once safety programs are in place, they cannot be evaluated to determine if they are effective and whether resources committed to safety are being used wisely. Data of this type are limited relevant to the fishing industry. Data that are available are not being used for systematic performance evaluation of current safety initiatives.

Major problems pertaining to safety data uncovered during this study include:

- inadequate casualty statistics and reporting;
- incompatibility among data bases;
- incomplete data on fishing vessel fleets and status;
- absence of demographic data on fishermen;
- absence of comprehensive and systematic monitoring and evaluation of safety performance of vessels, personnel, and equipment; and
- general nonavailability of occupational safety and personnel injury data.

The initial safety-improvement alternatives identified by the committee address these data deficiencies.

Improving Safety Data

Alternative 1: Establish a Comprehensive Statistical Data Collection, Analysis, and Utilization Program

This alternative envisions establishing a comprehensive safety administration program to research data needs, establish collection programs, and develop evaluation criteria. Ideally, it foresees modifying and upgrading data coding to permit correlation of data from all relevant data bases, starting with those maintained by the Coast Guard, federal agencies and operating units, the insurance industry, and other organizations maintaining useful information, and with vital records maintained by the states. Implementation issues include overhauling existing data collection and recording practices, developing taxonomies to guide data coding, arranging to exchange or share data among organizations, and determining resource requirements.

Alternative 2: Require Vessel Registration

Although all uninspected commercial fishing vessels are required to carry state numbers or be documented with the Coast Guard, there is no complete record of which vessels are actively engaged in commercial fishing. As a result, the size of the national fleet to which safety measures might apply can only be estimated. Furthermore, there is no capability to monitor safety performance, because vessels cannot be tracked across the various data bases. This alternative envisions that the Secretary of Transportation use discretionary authority provided by 46 U.S.C.A. §12501 to require information beneficial to law enforcement officials. This could include reporting of key physical characteristics, vessel usage, and current employment status as part of the mandatory vessel identification system. Such information could be extremely useful in assessing the infrastructure and support needed to improve safety.

This alternative could use existing federal and state registration infrastructures for administration, expand the data, and make vessel registration a prerequisite for operations. The information is essential to determine the scope of safety problems and resource needs for implementing other vessel-related alternatives, provide a more complete basis for correlating data to identify problem vessels, and evaluate the effectiveness of safety-improvement alternatives. It could also provide important information for fisheries management. The Coast Guard's ongoing vessel identification system (VIDS), under which each vessel in the United States would be required to have a distinct identification number, could be modified to accomplish this alternative.

Alternative 3: Require Professional Registration

This alternative envisions requiring a new merchant mariners document in 46 U.S.C.A. Chapter 73. In its simplest form, it could provide a way to determine the number of individuals engaged in commercial fishing. Individuals could be annually registered with a suitable industry or government organization without regard to professional qualifications. Certain demographic information could be obtained as part of registration to better understand the population engaged in the profession. This alternative could be applied to the entire fishing industry, such as in Alaska, where a crew license is required.

Licensing as a safety-improvement alternative is discussed in Chapter 5. Professional registration could also be adapted to serve as a rudimentary certification or licensing scheme. Traditionally, implementation of licensing in the U.S. maritime industry has employed grandfathering provisions. Professional registration could be employed as a first step toward certification or licensing by requiring a professional registration document or card as a precondition for work in any capacity aboard fishing vessels. The card could be provided without a test, but made revokable so that persons violating regulations (such as failure to report a marine casualty) could have their commercial fishing or vessel-operating privileges suspended or denied. As a progressive measure, renewing registration after a certain period could be made contingent on certification that certain prerequisites—perhaps entry-level training—have been acquired. If registration were used in these ways, measures could be developed for interim registration to accommodate filling crew vacancies on short notice.

This alternative is attractive in that invaluable census data could be acquired. Additionally, if adopted as a government program, the existing Coast Guard infrastructure extending to major fishing ports nationwide, with few exceptions (notably in Alaska), could be used to facilitate implementation on short notice with provision of suitable registration materials and equipment. Most Coast Guard shore stations are already equipped with standardized computer systems. Registration software could be developed for use on these systems. The Coast Guard has already researched card technology for potential use in its port safety and security program (USCG, undated). Technology developed or adapted for this use is potentially transferable to a registration system for fishermen.

SUMMARY

The fishing industry has a high incidence of sudden catastrophic loss of personnel and vessels. These events occur throughout the entire industry and are not isolated to any one segment. Certain areas are more prone to total vessel losses and vessel-related fatalities. However, exposure to vessel- and life-threatening situations occurs whether fishermen operate offshore, inshore, or on inland waters; in all environmental conditions; and on all sizes of vessels. The lack of data on the number of vessels, population at risk, and exposure factors makes development of precise casualty rates futile.

Because of the data limitations, it is difficult to determine the exact importance of any particular cause or type of incident or exactly how effective any particular improvement strategy might be. Because commercial fishing takes place in so many different environments, utilizing so many different kinds of vessels and gear, safety problems cannot be attributed to a few universal causes, nor can most incidents be easily attributed to a particular cause. Human, vessel, and environmental causes are all major contributing factors in casualties, but the data are extremely limited. Safety-strategy design, therefore, must incorporate other information beyond hard data—such as reasoning about likely causes and effects—and the vast anecdotal evidence available. Safety-improvement options include the following alternatives (sequentially numbered):

1. establish a comprehensive statistical data collection, analysis, and utilization program,
2. require vessel registration, and
3. require professional registration.

4

The Fishing Vessels

This study and the Commercial Fishing Industry Vessel Safety Act of 1988 (CFIVSA) arose from concern for people exposed to dangers on board fishing vessels. Danger is present during all phases of fishing operations—pretrip loading, transit to and from the fishing grounds, fishing, and unloading. Fishing vessels flood, founder, capsize, burn, go aground, collide, and break down. Ultimately, vessel loss or damage results, often accompanied by deaths or injuries. If forced to abandon ship, all on board may end up in the water or in a life raft. For those trapped on a sinking or capsized vessel, it can become a tomb.

The vessel as a working platform is the site of a variety of occupational accidents; fishermen fall, are knocked off, or become entangled in fishing gear and are pulled into the water. Fishermen increase their exposure to risk by the way they interact with the vessel, machinery, or fishing gear: for example, improperly operated winches, walking under suspended gear like crab pots, working on deck without protective clothing, or standing under a brailer full of fish. Although fishermen's actions and behavior are instrumental in the chain of events that cause accidents, not all accidents can be prevented solely by modifying their behavior. In some cases, accidents may be better prevented through vessel design modifications (National Research Council [NRC], 1985) and engineering and technical solutions. Therefore, understanding the vessel as a complex system of transport, domicile, workplace, and product storehouse and knowing what can happen to or on it are essential to modifying it to improve safety.

This chapter focuses on uninspected fishing vessels (see box, p. 83), i.e.,

Capsized fishing vessel *Melissa Chris*, Peril Strait, near Sitka, Alaska, August 18, 1988. (PAC Ed Moreth, *U.S. Coast Guard*)

that 99 percent of the U.S. fishing industry fleet subject to only very limited federal, industry, or self-imposed requirements governing design, construction, maintenance, or installed equipment. The nature and causes of vessel-related safety problems are discussed, along with safety-improvement alternatives. The analysis includes uninspected fishing industry vessels with combined catching and processing capabilities, but it also applies to fish tender vessels and non-industrial components of processing vessels. Vessels that transport only fish as general cargo were beyond the study's scope.

THE VESSEL AS AN INTEGRATED SYSTEM

A fishing vessel is a complex system in terms of function as well as engineering and technology. It is outfitted with propulsion and steering machinery; fishing gear; and deck, navigation, and communications equipment. Outfitting can range from austere—as in the case of small traditional boats like inshore lobster boats that still rely heavily on manual labor—to elaborate vessels with highly engineered, computer-controlled gear-handling systems and space-age electronic-navigation equipment like that found on large groundfish trawlers in Alaskan waters. Cargo space can range from simple fish boxes or bins to circulating seawater systems to flash freezers. All but the smaller vessels usually have cabins, pilot- or deckhouses, or similar shelter. Larger vessels have living quarters, galleys, and marine sanitation systems. These components form the complete engineering and technical system needed to catch, preserve, and transport fish.

Vessel Characteristics as a Safety Issue

Every vessel has a distinct "personality." No fishing vessel is identical to another, either physically or in handling characteristics. Even similarly designed vessels may be constructed from different materials; equipped with different propulsion or steering systems; modified while in service; or configured with different fishing gear, deck machinery, or electronics. Vessel loading conditions change dramatically as fuel and supplies are consumed, fish are harvested and stored, and gear is retrieved and stowed. This significantly affects a vessel's motion and handling, particularly stability (Adee, 1987). The harsh environment vessels operate in accelerates corrosion and wear. How these factors affect safety depends on how well the skipper and crew know their vessel—what it can and cannot do and how it reacts to varied loading and environmental conditions.

NATURE AND CAUSES OF FISHING VESSEL SAFETY PROBLEMS

Many fishing vessels are well designed, constructed, and maintained. There are some, however, whose seaworthiness is questionable. Unfortunately, there are no data to determine how many fishing vessels can be considered unseaworthy, either by design or by material condition. To obtain even a rough estimate would require a vessel-by-vessel physical review of design and construction documents (insofar as they were used and are still available) and inspection of material condition for a representative sample of the fishing fleet. Such an effort was beyond the scope of this study. Nevertheless, Coast Guard main casualty (CASMAIN) and search and rescue (SAR) data indicate that maintenance deficiencies involving hulls and fittings, propulsion equipment, and other systems affect all types of uninspected fishing vessels. Major stability problems are implicated by accident investigations (National Transportation Safety Board [NTSB], 1987) and anecdotal information. They can result from inadequacies in design, construction, or conversion; loading; or insufficient understanding of stability (see Dahle and Weerasekera, 1989; Hatfield, 1989; Plaza, 1989; Adee, 1987; U.S. Coast Guard [USCG], 1986b). In many cases, stability has not been considered in design. The apparent high incidence of workplace accidents suggests inadequately designed safety features in machinery, deck layouts, and fishing gear. Furthermore, many vessel parts are not designed; they are simply fabricated on the spot (American Society for Testing and Materials [ASTM], 1988; Miller and Miller, 1990).

Vessel Casualties

Fishing vessel accidents occur when the physical system—the vessel—fails, fishermen misuse or exceed its capabilities, or it is overwhelmed by external forces like rogue waves (see Grissim, 1990). Some factors can be altered to improve safety. This section draws on the results of National Transportation Safety Board (NTSB) investigations, the regional assessments, anecdotal information, and the committee's collective experience to assess inadequacies in or failures of engineering and technical systems leading to casualties. The nature and causes of casualties for documented vessels as recorded in CASMAIN are representative of those for comparable state-numbered vessels. Technical reasons for markedly lower vessel losses for state-numbered vessels are also discussed.

Flooding and Foundering (Sinking)

The principal vessel-related cause of flooding and foundering recorded in CASMAIN is failed material resulting in a breach of the vessel's hull, which frequently leads to stability problems. Flooding leading to foundering can be sudden, but the rate of water ingress sometimes provides time for the crew to control or correct the problem or evacuate. Flooding does not necessarily lead to sinking, however, notably in small fishing vessels with installed flotation. Representative flooding scenarios that can lead to foundering include the following:

- boarding green seas in heavy weather, thereby damaging or overwhelming closures or hatches on the weather deck (Dahle and Weerasekera, 1989; NTSB, 1987; National Fisherman Yearbook, 1982), or—in the case of open construction—swamping a vessel even in less severe conditions;
- in fair weather, shipping water through open, unattended machinery space doors, deck hatchways, or hull openings when the vessel is heeled over as the result of lifting an unexpectedly full net, snagging a trawl on the bottom, shifting cargo, the free surface effect of liquid in tanks or water on deck, or improper loading (Nalder, 1990; Dahle and Weerasekera, 1989; NTSB, 1987; National Fisherman Yearbook, 1982);
- neglecting minor hull leaks that open up in heavy weather and let water in beyond the capacity of the bilge pump (NTSB, 1987; Lesh, 1982);
- failing to detect hull leaks in unattended compartments with no alarm systems or one that is disabled or fails (NTSB, 1987; Taylor, 1985; Lesh, 1982);

- failing to detect improper, clogged, worn, or broken seawater piping systems, valves, flexible hoses, and pumps (Taylor, 1985; Nalder, 1990; Lesh, 1982); and
- breaching the hull as a result of collision or grounding.

Competent design and construction followed by periodic maintenance, presail and underway tests of equipment and alarm systems, and routine underway checks of unattended spaces and machinery are well-established, effective ways to prevent such scenarios.

Although few fishing vessels can survive engine room or lazarette flooding when loaded with fish, engineering measures can prevent sinking. They include the vessel's capability to pump out these two critical compartments, combined with working water-level alarms to alert operators to impending danger. Because deck loading or sea conditions can prevent access to the lazarette, pumping systems for this compartment could be configured to permit dewatering from another position on the vessel. Gravity drainage has not been effective in alleviating lazarette flooding.

Capsizings

Capsizings occur when vessels are made to operate in environmental conditions (e.g., wind, sea, and ice) or in ways (e.g., improper loading) that exceed their righting capability. Often, sudden overwhelming events with high fatality potential occur because of insufficient warning time to abandon ship. Persons on board can be trapped inside the hull, entangled in the rigging, or thrown into the water without personal survival equipment (see Chapter 6). Such situations are serious and life threatening.

Capsizings have been caused by a loss of stability resulting from undetected flooding; synchronous rolling; and disregard for, or ignorance of, intact stability (Adee, 1985, 1987; NTSB, 1987; Dahle and Weerasekera, 1989). Other common causes are overloading on deck and improper use of tanks, sometimes aggravated by icing (see Adee, 1987; Ball, 1978; Walker and Lodge, 1987). Vessels such as crabbers, which combine low freeboard and high deck loads, are particularly susceptible to accidents caused by inadequate intact stability. In other fisheries, vessels that experience high deck loads, such as scallopers and clammers, are also susceptible (McGuffey and Sainsbury, 1985; Sainsbury, 1985; Taylor, 1985). At this time, there is no consistently reliable, simplified test to substitute for an inclining experiment to determine static stability, or a vessel's ability to right itself (see Eberhardt, 1989). Many of these casualties result from inadequate or outdated stability information for a large portion of the fleet.

The effect of free surface liquids in tanks and fish holds can reduce intact stability. Such situations have led to a number of capsizings. Free liquid surfaces on board can also occur when water is trapped on deck. This is serious if

A good day's fishing and unusual cargo placement. Deck loading and cargo placement can affect a vessel's stability, reducing margins of safety during operation.

the weight and volume of water trapped high on the vessel causes it to heel over, flood, and capsize. Blocked or insufficiently sized freeing ports in bulwarks and other structural features can also trap water, making vessels particularly susceptible to free-surface-effect problems.

Stability is also affected when a vessel is modified or converted without considering lightship weight and center of gravity (NTSB, 1987; Adee, 1987). Installing refrigerated water or circulating seawater systems in a fish hold can significantly change an older vessel's intact stability, necessitating restrictions on loading and hold usage to compensate for free surface effects and added weight in seawater tanks.

Vessel size contributes to capsizing as well. A small vessel, despite good seamanship and material condition, can be more readily overwhelmed by heavy sea conditions (Canadian Coast Guard [CCG], 1987), as can almost any fishing vessel crossing a hazardous bar. Similarly, designing a vessel to comply with fishery management regulations for a primary fishery may compromise design for any secondary fisheries undertaken to make the operation profitable. Some fishery management regulations have been written with a principal dimension, particularly length, that is used as a cutoff point to limit carrying capacity. To comply with the rule, yet increase carrying capacity, a vessel's dimensions and proportions may be compromised.

Some of the circumstances reviewed above may be beyond any fisherman's ability to control. In other situations, however, a vessel operator may unwittingly push a vessel beyond its inherent capabilities. Thus, knowing a vessel's capabilities—including stability—is critical to safe operation. Although stability criteria for fishing vessels less than 79 feet long are not clearly defined, many good design features can be incorporated. For example, to minimize free surface effects, tanks can be sized and arranged to minimize the number that have free surfaces at any given time.

At all times, prudent seamanship and proper loading conditions are necessary. Despite the relatively low number of capsizings relative to other events, many vessels are frequently operated while exhibiting marginal stability as a result of loading. Such situations are potentially disastrous if conditions exceed the vessel or operator's capabilities.

Fires and Explosions

At sea, fire is even more dreaded than it is ashore. Fishermen faced with fire at sea can neither call for professional help nor run away from the danger. Short of abandoning ship in favor of a tiny liferaft, they must stay onboard and fight the fire themselves whether or not they have any training (Sabella, 1986).

An estimated 70 percent of fires and explosions are associated with vessel-related causes; approximately two-thirds of those leading to vessel losses occur in machinery spaces, and for fish processors, in dry cargo and refrigeration spaces. Other fires occur in the galley and living spaces. The causes are many and varied:

- improperly stowed combustible materials and highly flammable liquids (Sabella, 1986; Hollin and Middleton, 1989; USCG, 1983, 1986b; Taylor, 1985);
- buildup of hydrogen gas from batteries in confined, poorly ventilated spaces (Sabella, 1986; Hollin and Middleton, 1989);
- broken fuel, lube, or hydraulic oil lines spraying atomized oil onto hot surfaces (USCG, 1986b; Sabella, 1986; Hollin and Middleton, 1989; Taylor, 1985);
- faulty electrical systems, especially if standard home or industrial electrical equipment is used instead of those certified for marine use (USCG, 1986b; Sabella, 1986; Hollin and Middleton, 1989), and exposed lighting fixtures in contact with combustible material;
- attempts to repair fuel and lube oil systems at sea;
- open-flame heaters and radiators and improperly installed propane fuel systems (Sabella, 1986; Hollin and Middleton, 1989);
- unprotected polyurethane foam (used extensively for insulation, it ignites and becomes extremely toxic when exposed to open flame or high heat,

produces toxic gas, and is difficult to extinguish) (Sabella, 1986; Hollin and Middleton, 1989; USCG, 1983; Readings from . . . Alaska Seas and Coasts, 1979b); and

- flexible hoses and polyvinyl chloride (PVC) piping, which melt or burn, connected to through-hull fittings and left open or connected without an intervening valve, leading to uncontrolled flooding.

In many cases, fires are not detected in time to act effectively because there are no fire-detection systems or they fail to perform correctly.

Fire prevention, fighting, and safety at sea are well understood and written about (Sabella, 1986; Hollin and Middleton, 1989; Roberts, 1989; U.S. Maritime Administration [MARAD], 1979; Zamiar, 1982; Readings from . . . Alaska Seas and Coasts, 1979b). There is nothing unique to the engineering, technical, or operational aspects of commercial fishing that precludes adapting design, construction, operating, safety, and fire-fighting measures and techniques used for other types of commercial vessels (see USCG, 1986b; MARAD, 1979; Taylor, 1985). Methods and equipment are available to prevent, detect, or extinguish all but the most catastrophic fires and explosions. These include smoke detectors and portable or installed fire-fighting gear. However, except for required portable fire extinguishers, this equipment is not widely used.

Groundings

Groundings result principally from navigational errors, but also from propulsion or steering system failure; insufficient power to move off or away from a leeshore, shoal, or inlet during heavy weather; or inadequate anchors or anchor lines without enough scope for prevailing conditions (see CCG, 1987). Groundings frequently lead to flooding, and ultimately sinking or breakup.

Although largely attributed to human causes, there are engineering and technical factors that can reduce the risk of going aground:

- sufficient horsepower to permit maneuvering in restricted waters under adverse environmental conditions;
- suitable ground tackle to permit anchoring the vessel in any piloting waters during breakdowns;
- routine preventive maintenance to ensure that each system is operating at optimum level; and
- watertight bulkheads to prevent progressive flooding.

Collisions (Including Allisions)

Although collisions primarily result from human factors, engineering measures can mitigate the hazard. Propulsion and steering systems and navigational

The fishing vessel *Alaskan Monarch* lost steering while beset in ice off the Pribilof Islands, March 15, 1990. The vessel was hammered by walls of ice and water in shallows near shore and driven aground. After the Coast Guard cutter *Storis* was unable to get close enough to pass a towline, all fishermen aboard were rescued by a Coast Guard helicopter from Air Station Kodiak. (Norman Holm)

equipment must be adequate for the vessel's service. Prudent design that includes such features as a collision bulkhead forward and transverse watertight bulkheads can contain flooding that could lead to capsizing or foundering. As with engineering design options for groundings, structural measures like these have not been widely adopted in fishing vessels.

Material Failures

Main engine failure is the leading factor implicated in vessel loss and damage from casualties coded as material failures in CASMAIN data, followed by propulsion and steering components. Why failure occurs is poorly indicated in the data. The operating environment contributes to maintenance difficulties, as does the timeliness and adequacy of maintenance.

There are significant regional variations in reported cases of material failures; 43 percent of all casualties to documented fishing industry vessels in the North Atlantic and 51 percent along the West Coast were recorded as material failures. About 20 percent was normal in other regions. The implication is that many vessels that are not fully fit for service are operating nationwide.

The prevalence of material failures in casualties in the North Atlantic and West Coast could be related to:

- a high number of older vessels, which are more difficult to maintain;
- on the West Coast, permits assigned to vessels rather than individuals, which favors retention of old boats rather than new construction; and
- part-time use of vessels for fishing, use of antiquated fishing gear or redundant capacity that results in low earnings, and depressed economic conditions (see Pontecorvo, 1986; Taylor, 1985).

Each of these factors might influence how much attention is paid to vessel maintenance.

While neither CASMAIN nor SAR data can be directly correlated or full cause-and-effect relationships established, the implication of each data set is that many vessels have inadequacies in material condition that could lead to material failure and major casualties. Many of the safety problems that result could be mitigated by applying basic engineering practices for design and maintenance.

Workplace-Related Casualties

Occupational safety and health issues are part of everyday life aboard fishing vessels. Some of them have been addressed in manuals and periodicals, but in practice few fishermen have been exposed to formal occupational safety practices. The record of fatalities unrelated to vessel casualties and the apparent high incidence of personal injuries unrelated to vessel casualties indicates that the vessel as a workplace remains a major problem area. This is due to a combination of work practices and vessel or equipment design (ASTM, 1988; Goudey, 1986a,b; Hopper and Dean, 1989; see Amagai et al., 1989; Carbajosa, 1989; USCG, 1986b). Examples of possible causes that could lead to personal injuries and loss of life are shown in the box (Alaska Fisheries Safety Advisory Council, 1977; Readings from . . . Alaska Seas and Coasts, 1979a).

The distinction between the vessel as an operating unit and a workplace is often obscure. Functions and activities associated with the work of catching fish, not operating the vessel itself, could be considered occupational rather than vessel related (i.e., operational)—for example, a finger amputated in a winch or unguarded machinery. Sometimes the distinction is blurred. A fall on a slippery deck could be either occupational (i.e, due to the absence of nonskid surfaces on decks or failure to wear nonskid boots), operational (from maneuvers resulting in unanticipated vessel motion), or a combination of both. Unfortunately, the data are not available to determine the full extent of workplace injuries (Canada, Goverment of, 1988; CCG, 1987; Gray, 1987a,b,c, 1986). Human factors engineering (HFE) resource materials are available, but have not been evaluated for use in the commercial fishing industry. No data were developed to indicate they have been applied. Some occupational safety

EXAMPLES OF CAUSES OF PERSONAL INJURY AND LOSS OF LIFE

- Stressed rigging
- Loose lines or gear on deck
- Loose or swinging rigging
- Improper use of machinery
- Overloaded skiffs or dinghies
- Poorly located controls and brakes on equipment
- Leaving machinery controls unattended while machinery is operating
- Working on equipment while it is running
- Slick decks
- Inadequate handholds
- Inadequate lighting
- Inadequate ventilation

engineering has been accomplished—for example, machinery guarding suitable for the marine environment (Figure 4-1). It is limited, however, and tends to apply to individual vessels and corporate fleets. Some vessel safety manuals are available to guide safe workplace procedures (see Hollin and Middleton, 1989; CCG, 1987; Sabella, 1986). Personal protection equipment, such as high-traction boots and wire mesh gloves, are available (Freeman, 1990; Safety at Sea, 1989). Data are not available concerning the degree to which such items are used in the workplace. The nature of personal injuries—especially to extremities—suggests a need for systematic, industrywide use of safety equipment and in-depth consideration of workplace safety.

Technology as a Safety Issue

Technology advances, especially over the last three decades, have accelerated changes to design, construction or manufacture, and use of vessels, equipment, gear, and fish-preservation systems (see Fitzpatrick, 1989). These advances have been incorporated into existing and new fishing vessels when fishermen perceived the benefit to their operations (Browning, 1980; Dewees and Hawkes, 1988; Levine and McCay, 1987). For example, more-precise position fixing has permitted lobstering by small vessels farther offshore, and use of larger and heavier fishing gear. Adopting new technology like electronic navigation devices and computer-controlled net-handling systems has also enhanced many fishermen's operating skills.

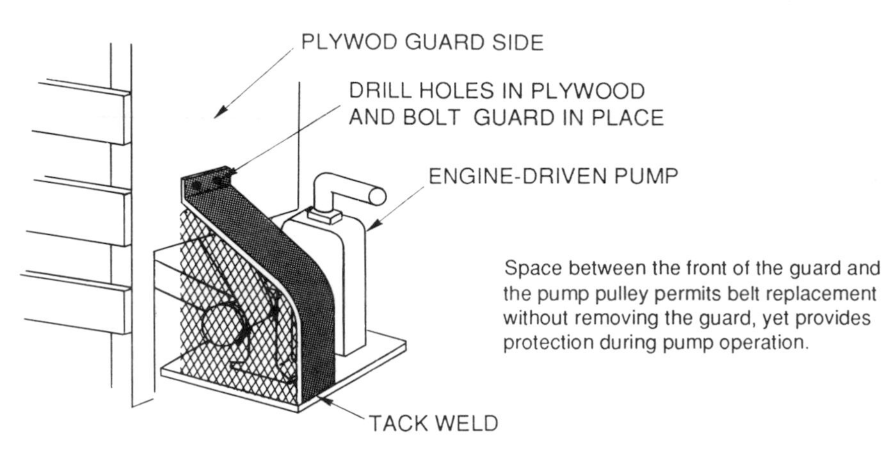

FIGURE 4-1 Example of machinery guarding for an engine-driven pump.

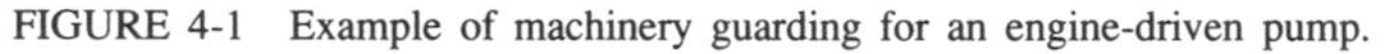

Using new technology or finding new applications for existing technology can have disadvantages, including increased risk-exposure associated with extended operations, however (deCarteret et al., 1980). For example:

- Experience operating newer hull forms and converted vessels and their idiosyncracies is more limited than it is with older boats whose shortcomings and responses are well known (see Gray, 1987b).
- High-tech systems such as loran or satellite navigation systems can fail at sea, but can be repaired only by shore-based technicians.
- Undetected failure of systems or their critical components (such as pumps circulating seawater to fish tanks) can lead to flooding, instability, foundering, or capsizing.
- Fishermen may become overly reliant on technologically advanced systems at the expense of maintaining adequate margins of safety (e.g., to compensate for increased risk-exposure) or traditional skills like basic piloting and navigation (deCarteret et al., 1980; Gray, 1987c).

Fishing Industry Perspectives on Vessel-Related Casualties

The committee consulted with a broad cross section of the fishing industry through the regional assessments it commissioned to obtain the industry's perception of vessel casualties and causes. The views varied among regions and, within regions, between those who fish and those who provide shoreside support. Flooding, foundering, fires, and explosions were generally cited as major casualties among regions, as were grounding and collisions. On the whole, this is consistent with the CASMAIN data, but within regions what constitutes a serious incident may differ between the fishermen and the statistics.

For example, on the West Coast, respondents cited foundering as the most serious type of casualty. But foundering accounted for only 7 percent of all casualties, while material failures were nearly 52 percent. The differences lie in the perception of severity of incidents. Material failures, although greater in number, tend to be viewed as significantly less threatening, even though an engine breaking down while crossing a bar could be extremely dangerous. With regard to safety-improvement options, perceived problems vary significantly from actual ones. Thus, effectively communicating problems to the population at risk in order to achieve workable solutions is critical (see NRC, 1989b).

With regard to causes of vessel-related casualties, the CASMAIN data and industry perspectives correlate well. Vessel maintenance inadequacies were weighted heavily in the regional assessments. Loading and stability were also indicated as significant factors. Improving the condition of the vessel and equipment were frequently cited as desirable approaches to improving safety.

DESIGN, CONSTRUCTION, AND MAINTENANCE

Features of a Seaworthy Fishing Vessel

Compensating for or designing dangers out of a vessel requires knowledge of what might constitute a seaworthy vessel. For this discussion, a seaworthy vessel is defined as one that is designed, constructed or converted, equipped, and operated commensurate with the conditions and service for which it was intended (see Nixon, 1986, for a discussion of unseaworthiness as a legal issue). Technical guidelines for designing, constructing, and outfitting seaworthy vessels are documented in Coast Guard publications, classification society guidelines, and similar sources (American Bureau of Shipping [ABS], 1989; USCG, 1986b; ASTM, 1988). A vessel fully ready for service would have the following adequate features:

- physical features—reserve buoyancy, static stability, the ability to remain afloat after partial flooding, watertight integrity, adequate drainage, adequate hull structural strength, noncombustible materials certified for marine use, and hard piping runs wherever feasible;
- operating equipment and machinery—propulsion, fuel, steering, ventilation, electrical, dewatering, and sufficient fire-fighting systems; navigation and running lights; mooring equipment and ground tackle; redundant fire and flooding alarms (especially for inaccessible or sometimes unattended spaces); and emergency electrical and propulsion systems;
- navigational equipment—charts of the right scale for the vessel's transit routes and operating areas, plotting tools (e.g., dividers, parallel ruler), a compass, marine radio capable of communicating with land-based SAR facilities, a depth sounder, a radar reflector, a loran, and a radar;

- survival equipment—personal flotation devices (PFDs), protective clothing with flotation and thermal protection in northern climates, covered life rafts, emergency position-indicating radio beacons (EPIRBs), portable radios, visual signaling devices (flares), and emergency rations; and
- workplace safety features—deck layout to minimize exposing personnel to the elements and fishing gear, winch controls arranged to prevent unattended operation, spooling guides, protective guarding of moving machinery, nonskid decks and working platforms, maximum pilothouse visibility, properly sized deck equipment and running rigging, fixed nonskid ladders with handrails, and section boards for cargo spaces.

In the United States, most uninspected fishing vessels are not required by law or regulation to meet most of these features. This will change significantly for some as the result of the CFIVSA.

State of Practice

In the ideal world, seaworthy fishing vessels would be those designed by naval architects specializing in the field, constructed by boatyards recognized for quality production, operated and maintained by experienced fishermen skilled in maintenance, and supported by technical specialists for high-technology equipment. The real world presents an altogether different picture. Vessel design, construction, conversion, operation, and maintenance for the most part are not subject to regulation. As a result, design and construction range from backyard boatbuilding techniques to those required for full compliance with classification society rules. Vessels range from those that should not be on the water to those that will, with proper maintenance, run with nearly clockwork precision throughout their service life. The results are predictable, as has been demonstrated by a continuing toll of vessels and personnel.

The nature and numbers of vessel losses, casualties, and associated accident reports implicate lack of design and poor construction in the sequence of events leading to some accidents. Furthermore, the fact that over 3,100 fishing vessels require assistance from the Coast Guard each year, 60 percent of them disabled and adrift, suggests that engineering practices are inadequate or not properly applied to the fishing industry. The significant incidence of failed material recorded in CASMAIN as a cause of marine casualties contradicts a commonly held perception in the fishing industry that fishermen will maintain a fully seaworthy vessel in order to catch fish. But a vessel can be taken to sea that is not fully fit for service and still successfully engage in fishing. It is apparent that some owners and operators maintain what is necessary to perform the vessel's primary function; others may compromise or delay maintenance because of economic factors, insufficient technical expertise, failure to detect wear or stress, and time pressures.

Design, Construction, and Outfitting Issues

Because commercial vessels may engage in interstate commerce, safety regulation falls to the federal government and is largely carried out by the Coast Guard, for example, federal regulations that pertain to design and construction of large commercial vessels with technical design and quality assurance reviews during construction provided by the Coast Guard. Ships like tankers and large passenger vessels have comprehensive technical and outfitting requirements and are subject to formal Coast Guard inspection. A major safety issue facing federal decision makers is whether inspection should be expanded to include uninspected fishing vessels. Background information on formal vessel inspection administered by the Coast Guard is presented in Appendix G.

Statutory and regulatory requirements governing the design, construction, and outfitting of uninspected fishing vessels are limited in scope and applicability. Existing regulations address fire-extinguishing equipment, backfire flame control, and hull ventilation (46 CFR Subchapter C). Small, noncommercial boats (principally recreational) under 20 feet in length converted for commercial fishing originally have to meet regulatory requirements for safe loading, including stability; safe powering, outboards only; flotation, fuel, and electrical system standards if gasoline powered; and ventilation. Fuel, electrical, and ventilation standards also apply to noncommercial vessels over 20 feet, which may incorporate some or all of the features required of the small boats (46 U.S.C.A. Chapter 43; 33 CFR Subchapter 5). If used commercially, warranties associated with the aforementioned requirements are nullified.

In some regions, many small boats originally designed for noncommercial purposes are used for commercial fishing. Those with built-in flotation do not necessarily sink when flooded or turned upside down and are thus potential survival platforms. This and the inshore nature of small-vessel operations may be one reason fewer vessel total losses and fatalities are recorded for small, state-numbered commercial fishing vessels.

The Coast Guard has proposed expanding requirements for engineering systems and equipment for all uninspected fishing vessels under the CFIVSA. But those built or significantly altered after the effective date of the regulations would also be required to meet operational stability requirements. Under the notice of proposed rulemaking (NPRM) (Federal Register, 1990), selecting an individual qualified to evaluate stability is the responsibility of the vessel owner.

The Coast Guard announced in the NPRM that it would continue its underway law enforcement boardings (Federal Register, 1990). This is the principal method used to determine whether an uninspected fishing vessel meets the applicable regulatory standards (see Appendix F). A dockside boarding program was being considered to aid in implementing the final rules (Federal Register, 1990). Commercial fishing vessels are the only category of commercial vessels routinely boarded while underway for safety enforcement, although boardings could be reduced if the Coast Guard accepts third-party certifications as suggested in the NPRM. However, the cost of technical review and quality assurance services, such as those associated with Coast Guard inspections to ensure compliance with federal requirements, would be borne by the owner.

Although there are few mandatory design, construction, or outfitting standards in the United States, there are comprehensive, voluntary (and in some countries, mandatory), international, national, and classification society standards (ABS, 1989; USCG, 1989a, 1986b; ASTM, 1988; International Maritime Organization [IMO], 1980, 1977, 1975b, 1966). In general, they cover hull structure and arrangements, stability, machinery and marine engineering systems, welding and materials, and detailed periodic survey requirements. Some include operational safety guidelines as well. The standards may be detailed or refer only to requirements (usually referred to as "rules") of ship classification societies. For example, stability for fishing vessels under 79 feet is not clearly defined, but standards do exist.

A few larger fishing vessels are designed and classed when there is incentive to do so, such as registration, financing, or insurance requirements. Classification society options may include plan approval and survey during construction, independent stability certification (including lightship weight determination), and full classification. Only a nominal number of fishing vessels are classed, although some have gone through intermediate steps. Some larger fishing vessels are designed to class standards, even though they may not be officially classed. This is largely because of:

- the cost of classing a vessel (as much as $20,000 for a 50-foot boat and $30,000 for a 125-foot vessel);
- continuing costs to maintain a vessel in class;
- limited requirements by lending institutions and underwriters for classing; and
- few, if any, requirements for a foreign voyage that might necessitate evidence of fitness.

Maintenance Issues

Except for basic safety equipment like fire extinguishers, there are no regulations governing vessel or equipment maintenance and no universal industry standards or certification programs to promote it, although up-to-date self-help manuals and guides are widely available (Hollin and Middleton, 1989; Sabella, 1986). Self-regulation varies in thoroughness and effect and depends on the knowledge and skills of the captain and crew. Voluntary self-regulation takes a long time to catch on, but owner liability cases have fostered its use. Some industry organizations and self-insurance groups have promoted systematic attention to vessel safety by publishing relevant standards or guidelines (W. A. Adler, Massachusetts Lobstermen's Association, Inc., personal communication,

1989; J. Costakes, Seafood Producers' Association, personal communication, 1989; Jones, 1987; Nixon et al., 1987). The net effect is that while some owners and operators hold themselves accountable for safety, no one is held strictly accountable for vessel fitness prior to operations.

Some technical support is available to fishermen. A marine survey is a physical examination of the vessel, equipment, and associated records by an independent third party. Considerable technical expertise is required to perform a competent examination (see Knox, 1990). Annual marine surveys are conducted on approximately 20 percent of documented and 10 percent of state-numbered fishing vessels (Federal Register, 1990), generally to satisfy insurance underwriters. As a rule, the larger the vessel and the farther it fishes offshore, the more likely it is to have insurance coverage and thus be surveyed (many smaller, undercapitalized boats and those in depressed fisheries operate without insurance, however). As with yachts, surveys can also be performed to assist prospective buyers in selecting a vessel, determine causes and costs of accidents, and assess whether a vessel will perform as expected (see Knox, 1990). A marine survey, absent observed deficiencies or corrected discrepancies, does not ensure that a vessel is fit for service. Surveys vary in thoroughness, depending on underwriting requirements and the surveyor's experience (Expert, 1990). In the near term, surveyors with technical experience relevant to fishing vessels are far fewer than are needed to replace Coast Guard compliance measures (Federal Register, 1990; Expert, 1990).

Another option is periodic boatyard maintenance, conducted where technical support may be available. This support also varies in quality and affordability.

Finding an Experienced Designer or Surveyor

Adequacy of design, construction, maintenance, and outfitting standards ultimately depends on the competence and integrity of the individuals providing technical services. So, choosing an experienced, reputable naval architect, builder, or marine surveyor is important. There is no institutional mechanism to help owners find such technical support, however. Professional trade associations and societies for naval architects, marine engineers, and marine surveyors have actively promoted fishing vessel safety and publish relevant information to their memberships. The four marine surveyor organizations in the United States set minimum professional standards for members (Expert, 1990). Thus, knowledgeable, experienced professionals, some of whom specialize in fishing industry vessels, can be found. Yet, it remains possible for novices to design, build, and survey uninspected fishing vessels and other marine vessels.

More detailed examination of this issue was beyond the scope of this study. However, mandatory vessel and equipment standards could help weed out individuals unable to deliver corresponding levels of support. National professional organizations could assist in building a pool of qualified technical personnel by providing training to support the fishing industry and establishing or expanding accreditation programs.

STRATEGIES TO IMPROVE VESSEL SAFETY

Safety can be improved by addressing problems associated with technology of the vessel and its equipment and human interactions with it. As discussed earlier in this chapter and in greater depth in Chapter 5, the human dimension in fishing safety is very important. Human behavior can be taken into account in the way vessels are designed, constructed, outfitted, and maintained. This is done either by designing potential human failings out of the system or by using technical systems as a medium for motivating or forcing behavioral changes. This section identifies specific safety-improvement approaches to vessel-related causes of fishing vessel casualties.

Improving Vessel Fitness for Service

Except for some basic federal regulatory requirements, there are no design, construction, outfitting, or maintenance standards for uninspected fishing industry vessels. This is expected to change as proposed federal regulations expand outfitting requirements for both state-numbered and documented fishing vessels. Additionally, major structural and equipment requirements could apply to certain documented fishing vessels (Federal Register, 1990). If adopted, the regulations would address many structural and equipment issues identified in this report. However, as the proposed regulations are written, vessel-design, construction, and material-condition issues would remain for the majority of the uninspected fishing fleet.

Major vessel-related problems uncovered during this study include:

- nonavailability or lack of adherence to structural guidelines, classification society rules, and similar standards during vessel design and construction or conversion that lead to structural or stability problems;
- general nonavailability of stability data for each vessel;
- inadequate material condition of vessels and equipment, especially machinery, alarm systems, and survival equipment (see Chapter 6);
- nonavailable or inadequate operating equipment, including bilge alarms and smoke detectors, bilge pumps, and fire-fighting systems;
- use of machinery and fishing gear with inadequate occupational safety features;
- inadequate personal occupational safety equipment; and
- inadequate or insufficient survival equipment (see Chapter 6).

Vessel-related problems could be corrected by establishing minimum structural and equipment design and maintenance standards; incorporating occupational safety features into vessel, deck layout, and equipment design; installing or providing selected equipment including occupational safety and survival equipment; determining and providing data on a vessel's operating and stability characteristics; and implementing programs to motivate or compel improvement to minimum standards in each of these areas. Some techniques could influence human behavior, such as forcing fishermen's attention and resources to maintaining the material condition of their vessels and equipment.

The following improvement alternatives continue the sequential numbering begun in Chapter 3.

Alternative 4: Establish Minimum Design, Structural, Stability, and Material Condition Standards

Virtually any hull form can be designed, built, and placed into service as a commercial fishing vessel without review of plans, construction techniques, or materials used and without determining its stability. Techniques used in other segments of the maritime industry to ensure a vessel's suitability for its intended service could be applied to fishing vessels, for example, standards for material condition and voluntary or mandatory certification or inspection programs. This alternative envisions that standards be developed, codified, and made mandatory for new construction and conversions and that levels of acceptable material condition be retroactively established for existing vessels and equipment. These standards could be adapted from guidelines published by IMO or existing voluntary guidelines (e.g., *Coast Guard Navigation and Vessel Inspection Circular 5-86* [USCG, 1986b], *American Bureau of Shipping Guide for Building and Classing Fishing Vessels* [ABS, 1989], and *Standard Practice for Human Engineering Design for Marine Systems, Equipment, and Facilities* [ASTM, 1988]). To some degree, standards are addressed in proposed rulemaking required by the CFIVSA. Developing standards is an element of most of the alternatives addressing vessel-related problems, because the standards form a baseline for measuring conformance or compliance. A thorough benefit-cost analysis would be needed to determine the economic feasibility of imposing standards where few or none existed before.

Alternative 5: Expand Equipment Requirements

Most fishing vessels are not required to be outfitted with equipment, such as automatic alarms, that would mitigate risk or increase timely detection of unsafe conditions. The problems could be dealt with by expanding federal requirements to carry or install equipment to include additional safety features. This alternative builds on existing regulations requiring installation or carriage of certain equipment aboard uninspected fishing vessels. It is already being implemented under a federal rulemaking mandated by the CFIVSA. Basic issues are:

- What equipment will be required?
- In what quantity?
- On which vessels?

Alternative 6: Improve Human Factors Engineering of Vessels, Deck Layouts, and Machinery

Some work has been done in the fishing industry generally on a vessel-by-vessel basis to improve operational and workplace safety through the design of pilothouses and deck layouts, including lighting, machinery, and fishing gear (Goudey, 1986b; Hopper and Dean, 1989). For example, several of the newest longline vessels fishing in the North Pacific were designed to minimize personnel exposure to fishing gear and wind and sea conditions on deck (Buls, 1990; Griffen, 1990). In another example, a corporate fleet modified machinery guards for more-effective at-sea utilization (Lucas, 1985).

Innovations like the preceding longline example are best inserted during design and construction (or even conversion) rather than retrofitting (see USCG, 1989a; Miller and Miller, 1990). Unfortunately, the range of innovations has not been cataloged, and the degree to which human factors engineering, including application of ASTM standards, might improve safety in the fishing industry has yet to be determined.

Improving Vessel Safety Performance and Owner and Operator Accountability

Maintaining fishing vessels and equipment is a major problem within the industry. A safe vessel requires dedicated involvement by the owner, operator, and crew. Some fishermen may not know the mechanics of maintenance, so increasing their knowledge and skills could, for example, lessen engine breakdowns or gear failure under stress. However, training may not be enough. Current motivating methods (voluntary measures promoted by the Coast Guard and industry organizations) have not resulted in a well-maintained fleet. Widespread vessel condition problems indicated by CASMAIN and SAR data suggest that more-rigorous motivation measures are needed.

Alternative 7: Continue Compliance Examinations

This alternative, a standard Coast Guard practice, usually occurs while a vessel is fishing or in transit. Coast Guard compliance examinations constitute the principal method for exposing uninspected vessels to federal checks for mandated equipment and adherence to federal laws and regulations, including those pertaining to fisheries management conservation, marine pollution, and drugs. These examinations are conducted as underway operational boardings (USCG, 1986a). In 1989, about 6 percent of the uninspected fishing fleet was boarded (USCG, unpublished data, 1990). Whether these boardings were effective in improving safety aboard has not been evaluated by the Coast Guard. Locally intense, fishery-specific boarding programs, notably in southeastern Alaska, have demonstrated their ability to lower safety violations, but this has not been correlated with SAR and casualty data. Principal issues are whether compliance examinations could be used to motivate universal adherence to upgraded safety regulations, and to what degree. The committee's assessment of Coast Guard compliance examinations is in Appendix F.

Alternative 8: Require Self-Inspection

In the absence of effective voluntary self-inspection within the fishing industry, some level of compliance activity is needed to motivate improvement in safety. This alternative envisions *mandatory* self-inspection with audits and on-site spot checks by government or independent, accredited, third-party inspectors. Innovative checklists could be designed to lead the operator or crewman through self-inspection of the vessel and equipment prior to the fishing season or an extended voyage. The checklist would be validated by the captain or other responsible individual and retained on the vessel, with a copy provided to an auditor. Major problems would have to be corrected according to established criteria or the vessel could not be operated pending repair and reinspection, with enforcement through the audit process.

This alternative could provide the fishing industry with methods to improve the condition of vessels and equipment and concurrently build safety awareness at moderate cost and minimal inconvenience. For owners and operators who already maintain their vessels, the impact would be minimal. Issues affecting implementation include provision of authority, falsifying checklists; the "checklist mentality," that is, focusing on the checklist to the exclusion of other safety aspects; and auditing criteria, responsibilities, and infrastructure.

Alternative 9: Require Marine Surveys

A more thorough check of vessels and equipment could be done by requiring marine surveys, which are already common for insurance. A federal requirement could call for marine surveys to verify conformance with applicable standards (alternative 4) at specified intervals and corrective action for deficiencies. Because each vessel is distinct, marine surveys could also be required when a vessel is sold or a new master or operator takes over to ensure that the captain is familiar with the vessel's material condition before operating it.

A variation of this alternative could require a marine survey for vessels that suffer major casualties or are the subject of a SAR incident implicating vessel condition as a major contributing factor. This alternative could limit government involvement and the need for additional federal resources. However, more qualified marine surveyors would be required. Implementation costs would be borne principally by the industry and vessels.

Alternative 10: Require Load Lines

An undetermined number of fish tender and processing vessels 79 feet or longer are subject to load-line regulations. However, fishing vessels are specifically excluded by law from load-line requirements (46 U.S.C.A. §2101; USCG, 1990b). These regulations establish the minimum safe freeboard to which a vessel may be safely loaded to its limiting draft (46 U.S.C.A. §5104). Vessels subject to the regulations cannot be operated unless load lines have been assigned (46 U.S.C.A. §5103). Load-line surveys consider the hull and fittings of the vessel, hull strength, stability for all loading conditions, overboard drainage of deck water in heavy weather, and exterior protection for crew members (46 U.S.C.A. §5105). These regulations are directed toward merchant vessels whose hatches are secured and made watertight for the duration of the voyage. Vessels subject to the regulations cannot be operated unless load lines have been assigned.

This alternative would expand load-line requirements to fish tenders and processors currently grandfathered under existing law and to fishing vessels, where practical, to take advantage of annual inspections to ensure hull integrity and quality and water- and weathertight closures. The full benefit of load lines would not be possible for fishing industry vessels, since they must be opened at sea as part of normal operations (Kime, 1986). The American Bureau of Shipping assigns load lines under delegation of authority in 46 U.S.C.A. §5107, with costs of the load-line inspection and corrective actions borne by the vessel owner. Similarly, Det norske Veritas Classification (DnVC) has been authorized to provide load-line assignments for U.S.-flag uninspected fish processing vessels that are either unclassed or classed by DnVC (USCG, 1990a). Coast Guard involvement is limited to enforcement. An interim-load-line-enforcement program was implemented by the Coast Guard for fish catcher/processor vessels in July 1990 (USCG, 1990b).

Alternative 11: Require Vessel Classification

In this alternative, fishing vessels meeting certain thresholds could be required (as new fish processing vessels will be under the CFIVSA) to be designed, built, and constructed according to ABS or similar organizations' rules. Vessels constructed and maintained under these rules would exhibit a high level of fitness for service. This alternative could limit government involvement and resource needs, drawing instead on the private sector. Costs would be borne principally by affected parties. Costs could be significant, however, particularly for the individual owner/operator.

Alternative 12: Require Vessel Inspection

"Vessel inspection" as traditionally applied to commercial vessels and referred to in this report is the formal program conducted by the Coast Guard, introduced in Chapter 2 and described in Appendix G. Coast Guard inspection of merchant vessels is rigorous and thorough to ensure that vessels meet minimum standards of fitness for service. A Certificate of Inspection (COI) attests that fitness criteria have been met and promotes—but does not ensure—proper use or maintenance.

This alternative would extend the Coast Guard's vessel inspection program to uninspected commercial fishing vessels. The need for basic fitness of state-numbered fishing vessels is no different from that for documented vessels. The general nature and causes of vessel-related safety problems for vessels of comparable length and employment appear consistent. However, the scope of inspection could accommodate variations in vessel types, local operating conditions, nature of the fisheries, and other factors.

Implementing a Coast Guard program to inspect fishing vessels at a level comparable to that for merchant vessels could impose considerable infrastructure requirements on the agency and considerable costs to the industry. The Coast Guard does not have the budget or personnel to expand its existing infrastructure to implement an inspection program. As discussed in Chapter 1, the Coast Guard's 1971 study found that such a program is costly and could create economic hardships for many fishermen. The report recommended a modified inspection program that would combine periodic inspection of requisite items with an advisory service to owners of documented fishing vessels (USCG, 1971).

A significant policy issue is administrative responsibility. Should the Coast Guard exercise exclusive authority for ensuring the fitness of state-numbered commercial fishing vessels operating on federal waters? The fishing industry is a hybrid situation. Even federally documented fishing vessels are required to obtain state licenses or permits for state-controlled fisheries. Conceptually, inspection of state-numbered vessels could be similar to state motor-vehicle safety and pollution inspections, perhaps drawing on that infrastructure (inasmuch as fishermen who drive cars are already within the system) and the existing relationship between the Coast Guard and state boating administrators concerning recreational vessels (see 46 U.S.C.A. Chapter 131). Alternatively,

potential jurisdictional conflicts could be resolved by expanding federal documentation requirements to all commercial fishing vessels, thereby providing exclusive jurisdiction for fitness to the Coast Guard.

Inspection programs could be implemented by expanding the existing Coast Guard or state inspection infrastructures. Implementation issues for state involvement include determining whether to inspect state-numbered vessels, standardizing minimum requirements, establishing authority for such a program (this also applies to Coast Guard inspection), and funding the inspection infrastructure. Determining state interests was not within the scope of this study.

A variation of full vessel inspection could be a program that falls between full inspection and compliance examinations, perhaps with an advisory service similar to what the Coast Guard considered in 1971. This option, for example, could employ a checklist similar to that in alternative 8, which could be jointly prepared by the vessel owner and a Coast Guard inspector prior to operation. Once major discrepancies are corrected, the vessel could be issued a document or sticker as evidence of compliance for a specified period. Additional implementation issues include modifying Coast Guard boarding policy to discontinue underway safety checks, unless there are apparent deficiencies, and use of neutral third parties or state infrastructures. The Coast Guard is already considering some of these options (Federal Register, 1990).

Removing Vessels from Service

Alternative 13: Remove Unsafe, Inefficient, or Excess Vessels from Service

There has been limited federal or state authority to reduce excess harvesting capacity or eliminate unseaworthy vessels through vessel removal or retirement programs. Such programs are in wide use in some nations—Canada, Norway, and Japan, for example. In the United States, however, vessel removal in the form of a buyback has been used only in the Pacific salmon fishery during the early and mid-1980s; an Alaskan buyback program was declared invalid under the state constitution. The purpose was to reduce overcapitalization in Washington State resulting from federal Indian fishing-rights decisions, but there was no moratorium on new entrants (Jelvik, 1986; Rettig, 1986; Koch, 1985). This approach was eventually discontinued in favor of limiting entry through a licensing program. Alternative 13 envisions a way to place vessels into a standby status, if they are in good condition and fishery resources might subsequently allow reentry, or retire them permanently to benefit safety as a primary or secondary objective.

Some vessels are not fit for service, as a result of either design, conversion or alteration, or material condition, while others may have become marginal producers through factors such as antiquated fishing gear. Presently, most fishing vessels are removed from service when they are uneconomical or so unseaworthy that no one will operate them commercially or when they are claimed by the sea. Other methods employed have included intentional, attempted, and alleged scuttling of vessels to collect insurance money (Lazarus, 1990a; Providence Journal, 1990; Letz, 1986; see Salit, 1989; Sullivan, 1984a,b,c,d; Clendinen, 1984).

Programs to remove excess harvesting capacity have generally been fishery-specific and have not necessarily precluded subsequent resale and use of the vessel in other fisheries or regions, which would merely shift capacity rather than eliminating it (see Mollet, 1986). Thus, the side effects of actions to reduce harvesting capacity need to be considered. One possible benefit of a vessel-removal program could be improved crew competency. As a fishery becomes overcapitalized, the returns are decreased and fishing effort is increased. The experienced crews go to the more profitable vessels. The least profitable vessels then are generally left to recruit inexperienced crews. As profitability is decreased, many vessels sail short-handed—i.e., a vessel ordinarily with a crew of four would sail with a crew of three. Removing the marginal producer could lead to a smaller but more experienced work force.

Limiting the size of the U.S. fishing fleet through a comprehensive removal program has not been attempted, and legislation would be required. This approach could reduce excess harvesting capacity and eliminate marginal producers and unseaworthy vessels from the fleet. The committee believes that this approach is worthy of consideration, but detailed examination is beyond the scope of this study. However, alternatives 8 through 12, as well as the Coast Guard's authority under the CFIVSA to terminate operation of "unsafe" vessels, have varying potential to force removal of unseaworthy vessels from service without government compensation.

SUMMARY

There are no engineering (as distinct from economic) impediments to producing a fit fishing vessel, the questions about clear definition of small-vessel stability notwithstanding. Most—if not all—engineering and technical issues can be compensated for somehow. Vessel design and maintenance are clearly very important, but vessel performance cannot be separated from human interaction with the vessel as an integrated system. Of known causes, vessel-related causes account for a significant portion of vessel casualties and approximately a third of fatalities among commercial fishermen. The factor most easily addressed through engineering design and operating practices, and the one that accounts for over 85 percent of known causes of vessel casualties, is the material condition of the vessel and its equipment. Inadequacies pertaining to stability can be addressed by adherence to IMO, classification society, Coast Guard, or similar stability guidelines during design for new construction or conversion, and by operating restrictions, followed by examination to ensure that a vessel is suited for its intended service. Stability criteria for fishing vessels under 79 feet in length are not completely developed. Techniques exist for determining intact stability of small fishing vessels at moderate cost, albeit in a static condition. Loading problems can be addressed through training and practicing good seamanship. Standard practices for human engineering design are published for marine systems and equipment, but have not been adapted or applied in the fishing industry. However, there are no universal mechanisms or procedures in place designed to ensure production, maintenance, or inspection of quality fishing vessels well suited for their intended service.

Vessel-related safety-improvement alternatives (continued from preceding chapters) that might be employed are:

4. establish minimum design, structural, stability, and material condition standards;
5. expand equipment requirements;
6. improve human engineering factors of vessels, deck layouts, and machinery;
7. continue compliance examinations;
8. require self-inspection;
9. require marine surveys;
10. require load lines;
11. require vessel classification;
12. require vessel inspection; and
13. remove unsafe, inefficient, or excess vessels from service.

5

The Fishermen

The role individual fishermen play in the chain of events leading to vessel casualties and fatalities has long been a concern in the fishing industry (Piche et al., 1987; Piche, 1985; Pizzo and Jaeger, 1974). The extent of that role is not sufficiently appreciated in many marine accident investigations because they focus on primary causes without exploring underlying factors: for example, whether the right decisions were made or qualification standards met, rather than the contribution of human behavior generally (National Research Council [NRC], 1981). By contrast, other injury-prevention research in the United States has concentrated more attention on the role of people rather than physical and product-oriented issues (NRC, 1985).

In marine casualties, terms like personnel error, human error, human causes, and personnel fault are not well defined. They generally refer to errors in judgment or acts of commission or omission leading to a casualty, and to ignorance and poor training (Dynamics Research Corporation, 1989; NRC, 1985, 1981, 1976). This report uses casualty terms consistent with the data and literature, but collectively the term "human factors" is preferred because it encompasses all human attributes bearing on safety.

Possible safety-improvement options related to human factors are identified and discussed in this chapter. Quantitative data alone are not conclusive, nor do they provide sufficient insight into human factors. However, the anecdotal evidence is abundant. That information, coupled with human factors research in other industries, has guided the committee's examination of this important area of concern.

HUMAN FACTORS AND VESSEL CASUALTIES

There is rarely a single cause of accidents. When people interact with machines, there is usually a chain of events involving more than one aspect of a system (U.S. Coast Guard [USCG], 1989a; American Society for Testing and Materials [ASTM], 1988; Dyer-Smith and De Bievre, 1988; NRC, 1985, 1981). It is a well-known fact among human factors engineering/ergonomics professionals that often what is attributed to "human error" in accident investigations is, in fact, "engineering error." That is, the engineering design of controls, displays, and/or workspace arrangements, etc., failed to adequately take into account human performance characteristics, capabilities, and limitations (see Miller and Miller, 1990; deCarteret et al., 1980). More in-depth investigations of accidents in other transportation-based industries (e.g., military and civil aviation, automobile/roadway) have shown a high percentage of these "human errors" to really be the result of inadequate human factors engineering design of the system. In fishing, cause and effect can be associated between the fishermen and their vessels, propulsion systems, deck machinery, fishing gear, navigational equipment, the operating environment, training and experience, or a combination of these factors.

Human causes of accidents include improper procedures, inexperience, poor judgment, carelessness, and navigational error (ASTM, 1988; Ecker, 1978; Esbensen et al., 1985). Also cited in major casualty reports and marine accident research are stress, fatigue, and boredom, which are critical to vessels at sea for prolonged periods or operating in congested ports and waterways (NRC, 1981; National Transportation Safety Board [NTSB], 1987; see Canadian Coast Guard, 1987).

Scope of Human Involvement

There are widely differing assessments of the role of fishermen in vessel accidents. Coast Guard data (1982 to 1987) implicating human factors as a primary or contributing cause were as low as 16 percent for search and rescue (SAR) data and 28 percent for main casualty (CASMAIN) data. The CASMAIN percentage was nearly 36 percent when secondary causes were included. These percentages, even with secondary cause data considered, are lower than might be expected. It should be noted that primary cause data identify known causes for only 55 percent of CASMAIN's 1982-1987 casualty records, and secondary cause data reflect known causes for only about 30 percent.

Some sources attribute 50 percent of all commercial vessel casualties (Dynamics Research Corporation, 1989; USCG, 1989a), and 70 to 80 percent of all marine accidents (USCG, 1989a; NRC, 1981, 1976) to human causes. For over 20 years, the American Hull Insurance Syndicate (AHIS) has consistently documented human failure as the predominant cause of casualties to the large commercial vessels it insures (AHIS, 1964-1986). The committee believes the CASMAIN data for fishing vessels do not reflect the true impact of human factors (see Piche, 1985), and that the 50 percent estimate is more realistic, with allowances for variation by types of incidents.

There are several reasons to support the belief that human causes have a higher impact than indicated by the data. First, injuries that do not involve vessel casualties or incapacitation often go unreported, because of reluctance of individuals to risk liability or jeopardize their insurability. The data are also limited because, thus far, reporting of occupational safety and health information is not required for most of the fishing industry and is not practiced voluntarily. Coast Guard investigators are trained to carefully examine performance of licensed personnel, but there are no performance standards or license requirements for over 99 percent of the uninspected fishing industry fleet. Finally, people maintain vessels, but there are no mandated material standards for uninspected fishing industry vessels upon which to gauge how well people perform this maintenance. The high incidence of failure of hulls, propulsion systems, machinery, and other equipment recorded in CASMAIN and SAR data suggest that many owners and operators are deficient in maintaining vessels and equipment. Extensive anecdotal information also supports the contention that lack of knowledge or skills, neglect, carelessness, and inattention affect maintenance.

Geographic Distribution of Personnel Errors

Although fishermen commonly consider inshore operations significantly less dangerous than offshore fishing, SAR data indicate otherwise. "Personnel error" as the primary cause of fishing vessel incidents appears disproportionately higher in inshore and inland waters than offshore. This distribution is somewhat higher than for fishing vessel SAR cases generally (see Chapter 3), but the data do not reveal why personnel error is more prevalent inshore. A partial explanation is that in shoal waters, narrow channels, and inlets, knowledge of winds and currents is essential. Navigation and boat-handling skills and the vessel's capabilities are apt to be severely tested. Yet, the method too often employed is trial and error.

HUMAN CAUSES OF ACCIDENTS

In this study, behavioral factors were categorized to guide discussion of the human dimension with members of the fishing industry. The categories appearing frequently in marine casualty literature and data were:

- fatigue/stress;
- improper or inadequate procedures (including inadequate or unsafe loading/stability practices and inadequate watchkeeping);

- improper maintenance;
- inattention (including carelessness);
- inadequate human engineering in design;
- inadequate physical condition;
- incapacitation through use of alcohol and drugs;
- inexperience (including inadequate knowledge and skills and insufficient familiarity with the vessel or fishing activity);
- judgmental errors (including faulty decision making and risk taking);
- navigational/operator error (including inexperience and errors in judgment);
- neglect (including willful negligence);
- personnel relationships; and
- working conditions.

Few of these terms are precisely defined. In many cases they fail to describe specific behavioral acts. They usually represent subjective value judgments by the reporter or investigator of marine casualties (see NRC, 1981). Nevertheless, they represent common characterizations of the role of people in marine casualties.

Although certain categories are more dominant, cause-and-effect relationships are varied and complex. CASMAIN data were analyzed to determine which behavioral factors were indicated as causes of marine casualties in the fishing industry. It was reasoned that certain human causes might appear more frequently and, if so, could be targeted for improvement. This approach proved feasible for collisions, groundings, and material failures, but the data were insufficient for other casualties beyond general characterizations. Generally, the most commonly reported were operator error, judgmental error, improper procedures, and improper maintenance. Human causes recorded in CASMAIN for documented vessels are believed to reflect the human causes of accidents involving state-numbered fishing vessels for similar incidents (based on similarities in SAR data).

Grounding

Operator error was the principal cause of 323 of 691 groundings attributed to human causes (Figure 5-1). Human factors were the principal cause of 81 percent of all groundings; however, there are significant regional differences—34 percent of the South Atlantic and 25 percent of Alaskan vessel casualties. These rates were disproportionately high, which suggests that fishermen in these areas encounter navigational problems in restricted waters.

Collisions

Improper procedures, principally failure to keep a proper watch or some other error, were the principal human causes of 80 percent of collisions (Figure 5-2). Human factors were the principal cause of 69 percent of all collisions. Thirty-eight percent occurred in the Gulf Coast and 21 percent in the North Atlantic. Twenty-six percent of all Gulf Coast vessel casualties were collisions, more than double the rate in other regions. The implication is that many Gulf Coast fishermen encounter navigational problems in waters congested by a high number of vessels and offshore platforms.

Material Failures

Human factors were recorded as the primary cause of less than 10 percent of material failure incidents. Of these, however, over 70 percent were attributed to improper maintenance. This dimension is symptomatic of human behavior but is not a behavioral act per se. The data provide no indication as to why maintenance may have been improper. Collectively, all other human causes account for less than 3 percent of the recorded primary causes of material failures.

Falls Overboard/Disappearances

Information concerning falls overboard and disappearances is limited, but the numbers verify that they account for the most fatalities; 26 percent of all deaths were attributed to known cases of falling into the water and 35 percent

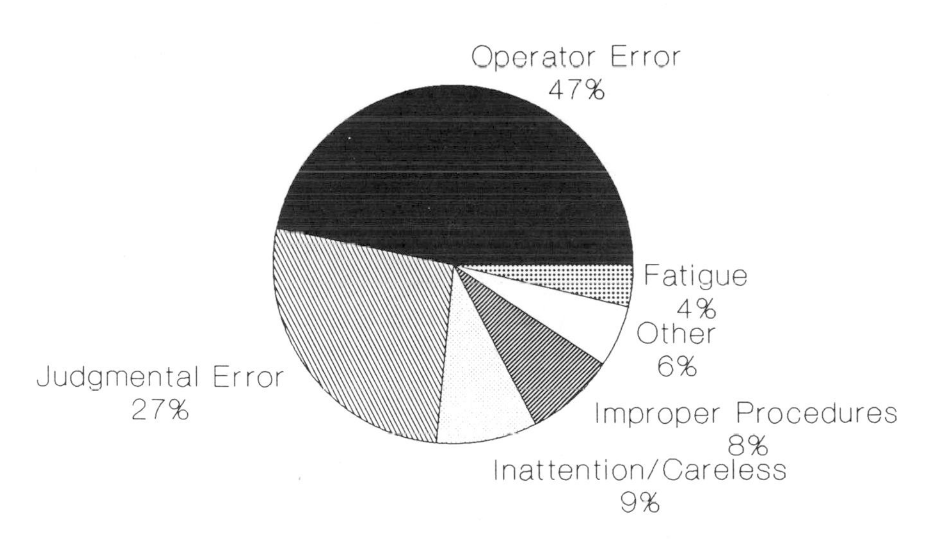

FIGURE 5-1 Human causes of groundings, CASMAIN data, 1982-1987.

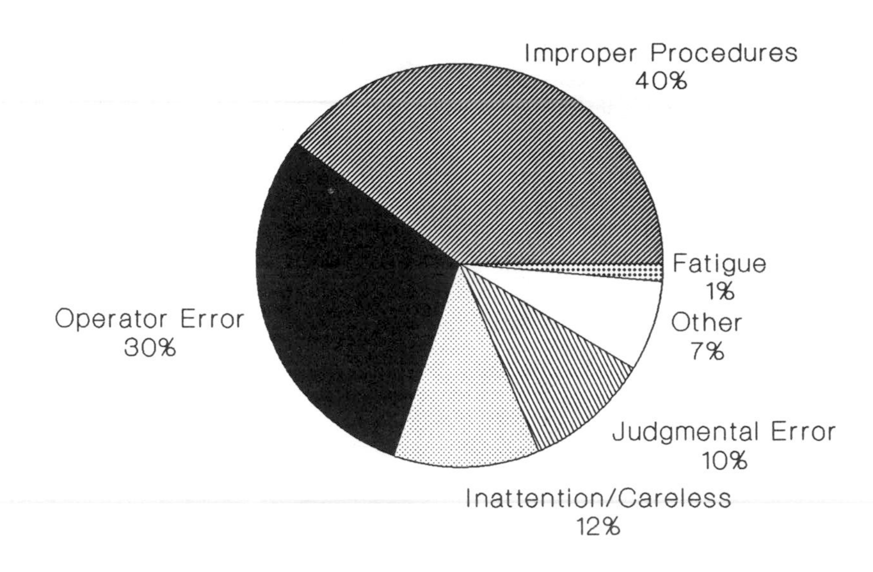

FIGURE 5-2 Human causes of collisions, CASMAIN data, 1982-1987.

to disappearances. Human causes associated with falls and disappearances are inexperience, inattention, judgmental error, neglect, and working conditions. Most fishermen do not routinely wear protective clothing with inherent flotation or other lifesaving devices while working aboard vessels, and these items are often inaccessible or unavailable in an emergency. Failure to use equipment of these types is infrequently recorded in CASMAIN data as the primary cause of a fatality. However, accident investigations and anecdotal information substantiate that nonuse of safety and survival equipment plays a major role in many fishing industry fatalities (see Chapter 6).

Injuries in the Workplace

So few data are available on injuries that the full scope of their frequency, cause, and severity is unknown, although their numbers are substantial (see Nixon and Fairfield, 1986; Alaska Department of Labor, 1988). CASMAIN data attribute about 39 percent of injuries not related to vessel casualties to human factors—the majority of them unsafe practices or movement about the vessel (see Table E-5, Appendix E).

LITERATURE AND RESEARCH

Literature addressing human factors in the fishing industry is scarce (see Dynamics Research Corporation, 1989; Esbensen et al., 1985; NRC, 1981,

1976). There has been no behavioral analysis of the maritime sector that breaks down operational problems into component parts for which human factors literature is available and then reassembles a composite view (see NRC, 1981). However, principles of human behavior related to safety have been published for consideration in the design of marine systems, equipment, and facilities (ASTM, 1988). There is no indication that ASTM human engineering design standards have been applied to fishing industry vessels.

The work of sociologists and anthropologists forms the largest body of literature about fishermen and their environment. Background papers on social organization and culture were prepared for this study (Gale, 1990; Maiolo, 1990), and a literature review was conducted (fishermen are characterized in Chapter 2). The other principal sources were NTSB accident reports and analyses (see NTSB, 1987) and preliminary results of Coast Guard-sponsored human factors research (see Dynamics Research Corporation, 1989; USCG, 1989a). This information was viewed in the context of operating conditions, platforms, and working conditions found in the fishing industry.

General Human Factors Research

Despite the paucity of research specific to the fishing industry, research on general human factors—such as stress in the workplace—may apply. Stress is of particular interest. It is a psychological or physical reaction to environmental factors such as noise, vibration, heat or cold, anxiety about achieving goals, drug use, interruption or disruption of sleep, and job performance (Hockey, 1986). For example, fatigue resulting from disruption or deprivation of sleep can lessen vigilance for certain tasks (Hockey, 1986). The usual effect of on-the-job stress is diminished performance, but severe stress may also result in health problems and physiological changes (Steiner, 1987).

Since fatigue is known to be a cause in some marine accidents, its causal relationship to fishing vessel safety is of interest. Fishermen routinely work 24-96 hours or more with little or no sleep (see Appendix H; Shafer, 1990a,b; Thompson, 1990; Steiner, 1987). CASMAIN data for marine casualties record fatigue as the primary cause in under 2 percent of human-related collisions and under 4 percent of groundings. Fatigue was not recorded as the primary cause in other categories that show the nature of incidents. The role of fatigue as a secondary cause is poorly understood in the context of commercial fishing, but is thought to be more important than indicated by the data.

INDUSTRY ISSUES

Expert accounts were contributed to this study concerning the inherent safety-related operational hazards and practical safety-improvement techniques in crab, troll, shrimp, and longline fishing. (The narrative on longline fishing is included in Appendix H.) These materials show the complex interaction of human factors and safety during fishing vessel transits and operations. The vessel captains who contributed accounts have incorporated safety into their operating procedures; thus, the material tends to reflect the upper end of the safety scale.

Onboard Conditions

The following conditions have an impact on human-factor-related safety problems on fishing vessels:

- employment without proof of professional competence;
- employment without screening of physical condition;
- operation of vessels without certification of professional competency, area knowledge, or familiarity with the vessel, equipment, or fishing gear;
- nature of employer/employee relationships (i.e., share partnership) resulting in lack of clear accountability for occupational safety and health aboard a fishing vessel;
- inadequate engineering of human factors and occupational safety in the design of machinery, fishing gear, and other equipment;
- general absence of systematic measures or programs to provide a safe workplace;
- continuous exposure to high risk during transit and while fishing;
- long, unregulated hours, often under severe conditions; and
- economic conditions that affect operational decisions and the level of risk exposure.

These considerations vary greatly by vessel, operator, owner, fisherman, fishery, and region (see Chapters 2 and 4). Adding to the lack of homogeneity is the population-at-risk factor, discussed in Chapters 2 and 3.

U.S. Fishermen's Perspectives

The perspectives provided in the regional assessments by leading professional fishermen and industry, management, and government leaders strongly implicated human factors as significant contributors to accidents. Different perspectives between fishermen and shoreside support personnel tended to vary in terms of degree rather than substance. Professional fishermen and other members of the industry implicated inexperience, inattention, and fatigue as the most likely contributors to safety problems. Operational procedures and navigational error were also indicated, but not to the same extent. Inadequate maintenance and loading conditions were also identified as major safety problems affected by human factors. While cause and effect cannot be determined with certainty, the most likely human factors in maintenance and loading are professional competence (e.g., skills, knowledge, and experience), inattention, neglect, and faulty decision making.

Fishermen nationwide have made clear their distrust of existing licensing programs. Yet, many have indicated willingness to participate in a licensing program that results in the selection of qualified and skilled personnel over those less qualified or skilled but more adept at taking written examinations. They are concerned that license preparation regimes can "program" individuals to pass license examinations by exclusively teaching examination questions and answers (published by the Coast Guard) without developing or corroborating acquisition of the basic skills and practical knowledge needed to operate effectively and safely. A similar theme for improving safety—emphasizing skill development over validating knowledge—was expressed by the majority of marine educators who responded to the committee's marine education questionnaire.

Members of the fishing industry generally urged that all safety-improvement strategies allow flexibility for local solutions, perhaps within a national or regional framework. Opportunities for local leadership in determining needs, developing and implementing practical strategies, setting realistic time frames, and considering economic conditions were not only urged, but indicated as essential to convincing fishermen that improvements would be in their best interest. These views are important, considering the lack of cohesive industry leadership and the findings of human factors research, which indicate limited utility in correlating personality traits and behavior factors with predicting injuries. Influencing behavioral change is most difficult for the groups at highest risk (NRC, 1985).

Canadian Fishing Vessel Safety Initiatives

The Canadian commercial fishing industry is similar to northern U.S. fisheries (gear types, stocks exploited, and environmental conditions). Safety problems examined by the Canadian government and fishing industry led them to recognize that safety awareness, basic professional knowledge, and skills were insufficient. Canadian investigations determined that existing wearable protective clothing was not well-suited to fishermen and was a factor in their decision to work on deck without flotation devices (see Chapter 6).

The Canadian Coast Guard (CCG) updated training standards for marine emergencies in 1988, conducted a safety awareness program communicating risks to fishermen in selected locations, and published a new fishermen's safety handbook incorporating voluntary self-inspection checklists and distributed it with annual fishermen's licenses. In addition, safety regulations have been proposed to require vessels under 150 gross tons to carry thermal decksuits specially developed for the fishing industry, and life rafts will be required regardless of vessel size or location after March 1991. A prototype life raft intended specifically for fishing vessels is being developed jointly by the CCG and industry. Existing inspection requirements are also scheduled for revision. (Canadian safety activities are reviewed in Appendix C.)

STRATEGIES FOR ADDRESSING HUMAN FACTORS

General Strategies and Human Behavior

Traditional resistance to past voluntary safety initiatives led the committee to consult injury-prevention research to gain insights on behavioral factors that might motivate safety performance and accountability. The research sets out three basic methods that may apply (NRC, 1985):

1. Persuade those at risk to voluntarily alter their behavior for increased self-protection. The Coast Guard's voluntary fishing vessel technical guidelines are an example (see USCG, 1986b).
2. Require behavior change by law or administrative rule. Existing examples are requirements to carry personal flotation devices (PFDs), and proposed new equipment requirements (see Federal Register, 1990).
3. Provide automatic protection through products or the work environment to mitigate the effects of human behavior. Self-activating emergency position-indicating radio beacons (EPIRBs), automatic water level alarms, automatic bilge pumps, and self-activating life rafts are examples found on some fishing vessels.

The research indicates that persuasion is the least effective, requiring behavioral changes has more of an effect, and providing automatic protection has the most effect. The fundamental reason is that members of high-risk groups (which include fishermen) "tend to be the hardest to influence with approaches that involve either voluntary or mandated changes in individual behavior" (NRC, 1985).

The unabated incidence of fatalities, injuries, and vessel casualties in the fishing industry suggests that fishermen most at risk have not been persuaded to voluntarily alter their behavior in the interests of self-protection. Many fishermen incorporate safety as a matter of good business. Unfortunately, many others appear not to. Therefore, the next step is to intervene to force behavioral change and require equipment designed to provide automatic protection.

The following considerations are also indicated (NRC, 1985):

- The effectiveness of safety strategies varies inversely with the extra cost and effort required to alter behavior.
- Laws and regulations intended to cause behavioral changes in individuals tend to be least effective among those exposed to the highest risk of injury.

• By itself, education has not proved to be an adequate preventative measure.

These findings have significant implications regarding the utility and implementation of specific alternatives for the fishing industry.

Because safety problems usually occur aboard vessels operating in isolation, away from even casual observation by law enforcement officials, any compliance activity requiring frequent on-scene observation or direct contact by external parties—such as the Coast Guard or the National Marine Fisheries Service (NMFS)—would be costly and its effectiveness difficult to determine (Blewett et al., 1985; Sutinen and Hennessey, 1986; see Sutinen et al., 1989a,b, for a detailed examination of fishermen's compliance behavior in relation to federally regulated fisheries).

Regarding education, specialized training of involved, interested individuals can substantially contribute to beneficial behavioral changes (McDowell Group, 1990, see Boehmer, 1989). The predominant theme expressed during this study and in the literature is improving basic nautical knowledge and skills. However, education has to be combined with translating it into effective usage. The common way to motivate effective use of knowledge and skills is a license or certificate that attests to competency and is subject to review, suspension, or revocation.

If compliance activities are expanded, they may come at the price of local leadership support, which is crucial to success (NRC, 1985; Pizzo and Jaeger, 1974). Many owners and owner/operators are already highly agitated over regulatory impositions. For example:

• changes to EPIRB regulations requiring high-cost replacement equipment;

• prospective long-term quota reductions in certain fisheries, such as king and Spanish mackerel (Stimpson, 1990);

• numerous law enforcement boardings by Coast Guard personnel on the fishing grounds (Sutinen et al., 1989a,b); and

• significant changes in harvesting practices, such as use of turtle excluder devices (TEDs) in shrimp nets (NRC, 1990; Williams, 1990; Cooper, 1989).

To avoid alienating local support, the preferred approach is to implement methods that motivate compliance rather than regulate it. Combining dockside inspection for mandatory requirements with advisory services to assist planning of vessel maintenance is one possible approach (Piche, 1985; USCG, 1971).

Specific Strategies and Alternatives

Major personnel-related problems uncovered during this study include the following:

- Perceptions of risk are inadequate, and awareness of safety procedures is limited, leading to risk taking, unsafe vessel operations, and unsafe work practices.
- There are weaknesses in basic nautical knowledge, practical skills in seamanship and maintenance, and motivation needed to safely operate and maintain fishing vessels, equipment, gear, and survival systems.
- There are weaknesses in knowledge and practical skills in basic navigation, vessel operating characteristics, and loading.
- There are weaknesses in knowledge, practical skills, and motivation in the use of workplace safety practices and safety and survival equipment.
- There are no formal professional qualification standards or requirements from entry level through senior positions for fishermen and most individuals in charge.

These problems could be corrected by building safety awareness and improving professional competency among the population at risk and supporting personnel. Elements of this approach could include communication programs; competency standards necessary to properly design, build, maintain, and operate fishing vessels; measures to encourage or compel acquisition of knowledge and skills needed to perform to standards; procedures for working on deck; emergency preparedness; measures to motivate safety accountability; and vessel manning and watchkeeping criteria.

Problems identified apply to operators of all uninspected fishing industry vessels. The committee found no evidence to suggest that fishermen aboard state-numbered vessels have a significantly better or worse safety record than their counterparts aboard small, federally documented vessels of similar hull design and gear configurations in similar service.

Building Safety Awareness

The committee found a general lack of enthusiasm for safety programs in the fishing industry. In allocating expenditures, items directly related to harvesting, transporting, and processing fish ordinarily take precedence. Survival equipment is usually a lower priority than vessels, gear, and equipment such as depth finders and loran. This is partly driven by costs, but individual perceptions of risk are also pivotal to motivating attention to safety (McCay et al., 1989; CCG, 1987; Dewees and Hawkes, 1988; Gray, 1987a; Levine and McCay, 1987).

Marine educators find widely varying perceptions about risks. Fishermen who voluntarily participate in safety training frequently were motivated by personal experiences or perception. Effectively communicating risks to fishermen is a central theme in other fishing industry safety studies (CCG, 1987; Pizzo and Jaeger, 1974), and increasing safety awareness needs to be a fundamental objective of a comprehensive strategy (McDowell Group, 1990).

Continuing the sequential numbering of alternatives in preceding chapters, the following safety-improvement alternatives are identified.

Alternative 14: Establish Risk Communication/Safety Awareness Programs

The fishing industry at large has a high-risk population, which is hard to influence with regard to changing human behavior. Therefore, the degree to which safety problems and solutions can be effectively communicated is crucial in establishing a cooperative environment to successfully implement other solutions. Failure to establish good faith and credibility leads to distrust and acrimony (NRC, 1989b), which are evident in the industry relative to federal regulation.

Risk communication could be employed to exchange safety information and opinions among affected individuals, groups, and institutions (see NRC, 1989b). This interactive process has already been started by congressional mandate through formation of the Commercial Fishing Industry Vessel Advisory Committee (CFIVAC) and its involvement in advising the Coast Guard on rulemaking. At the regional and port levels, this concept could be used to exchange information about hazards, experiences in dealing with them, and the associated economic impacts. It could also promote cooperation and build centralized leadership within the fishing industry to develop effective self-regulation. This would minimize further government intervention and limit the government's role to oversight and support. But to have maximum effect, risk communication must reach the population at risk—the fishermen.

This alternative expands existing practices, promotes mutual cooperation, and is relatively inexpensive.

Alternative 15: Publish and Distribute Safety Publications

A systematic program for publishing timely technical safety information to the industry could be established, building on existing, but mostly uncoordinated, efforts in this area. Communication options could include trade journals, government publications, and resource documents such as safety manuals for use in training programs. This could be a shared responsibility of industry, marine educators, and government agencies.

Such information is already published within the industry through many local, regional, and national forums. Some Coast Guard districts published special *Notices to Mariners* or digests in the past as safety guides for fishermen (see USCG, 1986c). These special publications appear to have been discontinued when regional fishing vessel safety manuals and voluntary safety guidelines were developed. Manuals cannot report day-to-day events, however. Currently, there is no periodical distributed to the fishing industry that the government uses as a forum to keep fishermen apprised of a broad range of safety issues. The role of trade journals and periodicals could be expanded in this area.

Relevant resource documents are among the infrastructural needs for implementation of safety alternatives. The committee's literature search uncovered a wide variety of written materials relevant to fishing vessel safety and resources for training and educational programs. However, resource documents have not been researched and cataloged for the benefit of training organizations and policy analysts. Potential exists for meaningful research in this area.

Improving Professional Competency

A prevalent theme of this study is that fishermen, even though skilled at catching fish, often lack the formal nautical skills and knowledge for safe operations. For example, many fishermen can operate electronic navigation equipment, but may not be fully proficient in basic chart navigation or principles of seamanship (see Gray, 1987c). The anecdotal information pointed to inattention or neglect of basics leading to accidents, such as failure to properly secure lazarette hatches leading to flooding, failure to wear PFDs while working on deck, and general unfamiliarity with survival equipment (see Chapter 6).

Optimum knowledge and skill levels to overcome these deficiencies have not been analyzed because of inadequate data, but knowledge and skills required to operate aboard vessels in other commercial maritime sectors could be adapted to the fishing industry. The knowledge and skills required to pass Coast Guard examinations and obtain licenses to operate uninspected towing vessels, uninspected passenger vessels for hire carrying fewer than six passengers, and uninspected fishing vessels over 200 gross tons have been accepted within the maritime community as legitimate. The areas of knowledge associated with these licenses are generally consistent with those used by other maritime nations and International Maritime Organization (IMO) training recommendations for fishing industry vessels (see Appendix C). Appropriate knowledge and skill levels (see box)—adjusted for operating environments by region, size or type of vessel, or numbers of personnel aboard—could reduce casualties in which inexperience, improper or inadequate operational procedures, judgmental errors, and navigational errors are contributing factors. Standardized procedures for vessel operations, deck work, and emergency drills could also increase competency.

Alternative 16: Require Emergency Preparedness Measures

This alternative would provide the means for onboard emergency preparedness education. Most fishermen involved in major vessel casualties are not prepared to effectively respond to emergency conditions. In other cases, the alarms and equipment designed to correct problems either were not available, did not work, failed to perform correctly, or were incorrectly used.

KNOWLEDGE AND SKILLS TYPICALLY REQUIRED OF LICENSED OPERATORS OF U.S. UNINSPECTED COMMERCIAL VESSELS

- Navigation and position determination
- Seamanship
- Watchkeeping
- Radar equipment[1,2]
- Compass—magnetic and gyro
- Meteorology and oceanography
- Ship maneuvering and handling
- Ship stability, construction, and damage control[1]
- Ship power plants (small-engine operation and maintenance)[3]
- Cargo handling and stowage[1,4,5]
- Fire prevention and fire fighting
- Emergency procedures
- Medical care (first aid)
- National maritime law
- Licensing and certification of seamen
- Shipboard management and training
- Ship's business
- Communications
- Lifesaving
- Search and rescue[6,7]

---

1 No requirement for operator, uninspected passenger vessels.
2 Required for ocean routes only.
3 No requirement for operator, uninspected towing vessels, oceans (domestic trade)/near coastal.
4 Required for barge operations.
5 No requirement for master or mate of uninspected fishing industry vessels, oceans/near coastal.
6 No requirement.
7 Recommended for fishing vessels by International Maritime Organization.

Basic knowledge and skills for responding to emergencies can be developed through education and training, and damage control and lifesaving equipment can be provided, maintained, and certified. But preparation is not a static condition. In an emergency, people are usually forced to act instinctively. Thus, on-site reviews of use and location of emergency equipment and frequent drills would enhance preparedness (McDowell Group, 1990; see Alaska Fisherman's Journal, 1989a). A side benefit is concurrent examination to identify deteriorating or defective emergency equipment. The Coast Guard has proposed a regulation that would require safety orientation, instructions, and emergency drills for fishing vessels, affixing responsibility for compliance on the master or the individual in charge.

Alternative 17: Develop and Promulgate Standard Operating Procedures

Every work effort, regardless of how small, develops operating procedures and design practice. They may be:

- fixed by custom, such as turning the steering wheel clockwise for right rudder;
- standardized by agreement, such as the direction valves are turned for opening and closing; or
- mandated by rules or regulations, such as nautical rules of the road.

Typically, onboard procedures vary as the result of vessel design and construction, installed equipment, deck layouts, fishing gear, and crew composition.

Some basic procedures could apply universally, such as the correct way to manually lift heavy loads, wear protective clothing, or stand watches. Procedures or guidelines covering operations and workplace safety could be developed from existing resource materials and published. Mandated use of such guidance could be difficult to enforce on isolated fishing grounds; therefore, strong local and regional industry leadership is a more logical way to promote this alternative, or assistance could be offered to develop guidelines, such as that previously provided by the Coast Guard for vessel safety and Sea Grant for safety training.

Alternative 18: Develop Competency Standards

This alternative envisions formally establishing competency standards for fishermen, operators, and masters of fishing vessels. Such standards should be robust enough to accommodate regional and fishery differences, yet contain the essential requirements to improve safety while minimizing the burden of implementation on the industry and Coast Guard. Standards are the prerequisite for practical safety training for vessel personnel and any "competency" procedures.

Competency standards for each level could include:

- for fishermen—rules of the road, safety and survival, seamanship, correctable vision, normal color vision, basic first aid, and basic fire fighting;
- for operators—advanced first aid—including CPR—navigation, and 1 year's experience; and

- for masters—advanced navigation, stability, advanced fire fighting, advanced seamanship, fishing vessel regulations, and weather forecasting.

Alternative 19: Promote Education and Training

Voluntary education and training could provide the means to acquire the knowledge and skills appropriate for competent fishermen, operators, and masters. This alternative suggests that aggressive marketing of education and training programs would provide the incentive for commercial fishing personnel to capitalize on such opportunities to increase their knowledge and skills.

No more than 10 percent of active fishermen are believed to have received such training; however, many fishermen are awaiting regulatory action by the Coast Guard concerning knowledge and skill levels prior to seeking training. Such actions, as presently contemplated by the Coast Guard, may reverse low demand experienced in current training capacity.

This alternative would need knowledge and skill standards and education and training regimes for determining training effectiveness. Further infrastructure development needed includes better collection and organization of vessel and personnel casualty data to establish a baseline by which to evaluate training effectiveness within the fishing industry, analyses on the effectiveness of different types of training (i.e., hands-on versus lecture versus videotapes), analysis of the effectiveness of different testing methodologies, curricula certification, instructor qualifications, participant certification, and adequate funding from present and alternative sources. Analysis of the effects of education and training on operator competency in other high risk industries, such as agriculture and trucking, might provide valuable insight for improving skill development programs for fishermen.

Alternative 20: Require Education and Training with Certification

This alternative would require education and training of all fishermen that would lead to professional certification or credentials. A certificate or credential implies a prescribed instructional program, certified instructors, and renewal and revocation processes. The training program could be developed jointly by industry, the training sector, and Coast Guard representatives and based on the competence or skills required for work in the industry.

Training could be practical and hands-on in nature to foster development of both skills and knowledge, and consider varying levels of competency based on vessel size and professional responsibility (crewman, operators). Potential guidelines have been put forth by the CFIVAC.

To ensure uniformity and quality, a basic standardized program would allow for regional considerations. Certification for instructors could also be implemented. Both the program and instructor certification could be monitored and evaluated by a national oversight group, such as the Coast Guard or Maritime Administration (MARAD). Consideration of past service and safety training (a grandfather clause) could be worked into any training scheme. Curricula needs indicated by accident investigations, CASMAIN data, and other materials include courses in personal safety, navigation, fire fighting, first aid, and safety equipment procedures.

Stable and continuing funding would be needed for development and initial implementation until programs could be financed by program revenues from the industry.

Alternative 21: Require Licensing

In this alternative, only individuals who hold a license would be authorized to occupy higher positions of responsibility and authority aboard a commercial fishing vessel. Indeed, the CFIVSA required the Coast Guard to provide Congress with a plan to license operators of documented fishing industry vessels. Marine licensing is a way to fix responsibility by limiting vessel operation to those who meet certain criteria. Licensing attests that established criteria have been met. In current marine licensing practices, this usually involves some form of written or oral examination after prerequisites (e.g., first aid training) have been satisfied through means such as required training and practical demonstrations. The concept is that licensed personnel would have the knowledge and incentive through improved accountability for their actions to be safer operators than nonlicensed personnel and would be involved in fewer marine casualties or incidents. This concept appears valid based on general safety trends, but has not been conclusively demonstrated through statistically valid analysis of marine casualty data or benefit-cost analysis.

The main difference between this alternative and requiring professional education and training with certification (alternative 20) is the means of meeting the criteria and administering the licensing program. In this alternative, satisfactory completion of standardized, formal examinations could be required before the licensing authority (Coast Guard) would issue a license. This alternative would, in effect, expand the Coast Guard's licensing program to include operators of uninspected commercial fishing vessels under 200 gross tons.

The applicant could be required to satisfy various prerequisites before sitting for the exam. Prerequisites might include, for example, documented experience or sea time, or satisfactory completion of approved training courses on topics such as first aid, CPR, radar, and seamanship. Mandatory prerequisite education and training would provide consistency of knowledge among vessel operators.

Traditionally, examinations have been used to attest to an individual's knowledge. An option exists that could reduce the need for additional licensing infrastructure and be more palatable to fishermen while still providing a means of attesting to acquisition of practical skills and knowledge. The licensing authority could accept certificates issued upon satisfactory completion of prerequisites from accredited sources in lieu of customary examinations.

The committee distinguishes between federally documented and state-numbered vessels, but finds no reason to treat operators of state-numbered fishing vessels differently from those of documented vessels with regard to basic knowledge and skills. However, the emphasis would vary by the nature of fishing activity, associated safety records, and costs versus expected benefits. For example, further examination of small-scale fishing operations on sheltered waters is needed to determine whether licensing of fishermen would have practical utility at this level, or whether requirements that might be developed at the state level for recreational boaters would suffice.

There is a significant policy issue concerning responsibility for licensing administration. Traditionally, the Coast Guard has exercised exclusive authority for licensing operators of commercial vessels operating on federal waters. But, as discussed in alternative 12, the fishing industry is a hybrid situation. Even fishermen aboard federally documented fishing vessels are required to obtain state licenses or permits for state-controlled fisheries. Conceptually, licensing of operators of state-numbered vessels could be similar to motor vehicle licensing, perhaps drawing on that infrastructure. Licensing programs could be implemented by expanding either the existing Coast Guard or state licensing infrastructure. If licensing were chosen and state licensing adopted, implementation issues include standardization of minimum requirements and license reciprocity. Alternatively, federal licensing requirements could be expanded to all commercial fishing vessels to establish exclusive operator licensing jurisdiction with the Coast Guard.

By combining professional registration (alternative 3) and grandfathering as the initial basis for issuing licenses, a licensing program could be implemented in the near term to improve accountability while the training infrastructure is being developed. License renewal could be contingent upon subsequent completion of training and examination requirements. Additional implementation issues include infrastructure requirements; funding, including cost-sharing with the states; license categories, based on position, vessel type, vessel size, number of personnel, etc.; competency standards; physical condition (e.g., correctable vision, normal color vision); and criteria for revoking, suspending, and renewing licenses.

Manning and Watchkeeping

Fatigue can degrade performance of navigational responsibilities and deck work. Proper watchkeeping during vessel operation, in addition to knowledge and skills, also demands alertness. Fatigue can slow reaction times, cause people to fall asleep while on wheel watch, and lessen the attentiveness needed to perform safely around machinery and gear. In addition to building safety awareness about the dangers of fatigue, it might be possible to change manning practices to provide checks and balances, such as maintaining two crewmen on wheel watch on rigorous trips (R. Jacobson, personal communication, 1990).

Alternative 22: Establish Vessel Manning and Watchkeeping Criteria

Minimum manning standards and watchkeeping requirements for navigation are published in federal law and regulations for most uninspected commercial vessels, including uninspected fishing vessels over 200 gross tons (46 U.S.C.A. Chapter 81). This alternative expands existing requirements to include uninspected fishing vessels of all sizes. Variations on this theme are being considered by the Coast Guard as part of the congressional mandate to submit a licensing plan to Congress for documented vessels. The evidence suggests that safety problems that might lead to manning and watchkeeping requirements for documented fishing vessels apply equally to state-numbered vessels, with flexibility for degree of emphasis as discussed in alternative 21.

SUMMARY

Human failure in some form contributes to most fishing vessel casualties, fatalities, and injuries. If not the direct cause, human factors are an element in accidents and complicate implementation of safety improvement alternatives. Human factors frequently associated with marine casualties are inexperience, inattention, fatigue, judgmental errors, and navigational errors. Safety can be addressed through voluntary or mandatory programs or systems designed around the human element. Safety-improvement options (continued from preceding chapters) include these alternatives:

14. establish risk communication/safety awareness programs,
15. publish and distribute safety publications,
16. require emergency preparedness measures,
17. develop and promulgate standard operating procedures,
18. develop competency standards,
19. promote education and training,
20. require education and training with certification,
21. require licensing, and
22. establish vessel manning and watchkeeping criteria.

6

Safety and Survival

An accident occurs and a fishing vessel begins taking on water; within minutes it lists and capsizes, and the captain and crew end up in the water. In another accident, a boat rolls in heavy seas and a crewman on deck loses his footing on the slippery surface. In an instant he is overboard. These are typical tragic scenarios in which fishermen have died because they were not wearing protective clothing, available life jackets, or survival suits. In all but a few incidents, fishermen in correctly worn immersion suits survived their ordeal. The continuing yet preventable loss of life among fishermen calls for a special assessment to identify ways that equipment can improve survival.

The committee considered equipment availability, testing, approval, labeling, performance, and effectiveness; use patterns; federal requirements; consumer issues; survival system information and literature; and equipment developments in Canada. Personal protection devices receive extended treatment because of the high loss of life that results when they are not used. The assessment presented in this chapter reflects the committee's experience; discussions with representatives of the U.S. Coast Guard (USCG), Canadian Coast Guard (CCG), Underwriters Laboratories, and survival equipment manufacturers; and regional assessments. Discussions and explanations of technical issues are provided in Appendix I.

DATA AVAILABILITY

Although there is considerable information about survival equipment, it is not comprehensive. Data concerning the performance of safety and survival equipment were not available from the principal sources used for this study—the Coast Guard's main casualty (CASMAIN) and search and rescue (SAR) data bases—but they were useful in identifying the types and frequencies of events that might necessitate the use of safety and survival equipment. SAR data were also useful in identifying the geographic density of these events. The most notorious fishing vessel disasters have been systematically evaluated by the National Transportation Safety Board (NTSB). NTSB accident reports include information on availability, use, and effectiveness of safety equipment and lifesaving devices, but are available only for major casualties.

Experience from disasters at sea and hypothermia research have established the need for buoyant devices to prevent drowning and protection against hypothermia. These needs are well publicized by the government, fishing industry trade associations, periodicals, educators and vocational trainers, manufacturers and retailers of lifesaving equipment, and even a phone company (National Council of Fishing Vessel Safety and Insurance [NCFVSI], 1989; NYNEX Information Resources Company, 1989; Texas Shoreline, 1989; Hollin and Middleton, 1989; Sabella, 1986; USCG, 1986b; Shafer and Beemer, 1984; Finley 1982a,b; Johnson, 1982; Myers, 1982). The effectiveness of survival equipment in providing emergency buoyancy, maintaining dry environments, and conserving body heat has been evaluated under controlled conditions (Offshore Research Focus, 1989; Steinman and Kubilis, 1986; Finley, 1982b; Johnson, 1982).

THE NEED FOR SURVIVAL EQUIPMENT

Survival Situations

The data presented in Chapter 3 indicate that the fishing industry has a high incidence of sudden catastrophic loss of vessels, which frequently leads to fatalities. Although these events occur throughout the industry, some geographic areas are more prone to total loss of vessels and fatalities, which needs to be considered in forming safety strategies. Exposure to vessel- and life-threatening situations occurs year round, whether fishermen operate offshore, inshore, or on inland waters, and regardless of vessel size. Thus, all fishermen need suitable survival training and equipment. This information has not been effectively communicated to fishermen.

Effects of Exposure

The effects of hypothermia and exposure are generally well understood. Research and experience have demonstrated that staying out of the water and conserving body heat are very important to survival (see Mercy, 1990; Zanoni, 1988; Alaska Fisherman's Journal, 1988b; Finley, 1982a; Johnson, 1982).

Survivors from the fishing vessel *Rainy Dawn* huddle with crabpots after their vessel sank near Kodiak, Alaska, September 8, 1989. (Andy Hall, *Kodiak Daily Mirror)*

For example, a 1986 Coast Guard research project examined the effects of protective clothing and survivor location after abandonment on body core and skin temperatures. Significant differences were found between bodily cooling rates in the water and on rescue platforms: "Survivors maintain higher skin temperatures and slower cooling rates out of the water, even when exposed to continuous wind, spray, and waves, than when they remain immersed in rough seas" (Steinman and Kubilis, 1986). Manufacturers have incorporated this information into innovative designs for life rafts and protective clothing.

EVIDENCE OF SURVIVAL EQUIPMENT INADEQUACIES

The Fishermen's Response

Some vessel owners, operators, and crewmen, particularly those operating offshore in northern climates, have voluntarily invested in immersion suits and float coats with fixed or inflatable flotation (USCG, 1989b; see Hamilton, 1989; Johnson, 1982). Inshore and midshore (within approximately 20-25 nautical miles of the coast) fishermen appear less inclined to perceive the need for immersion suits, citing proximity to shore and shorter running times to and from the fishing grounds as factors that lessen their exposure to life-threatening events (see McCay et al., 1989). Protective clothing, such as Coast Guard-approved, Type V antiexposure coveralls (i.e., deck and work suits), is infrequently provided by fleet or vessel owners. Such clothing is not used routinely while working on deck or in skiffs or tenders, although their use appears to be increasing aboard some large factory trawlers. The most commonly used outerwear is oilskins and similar waterproof rainwear, which do not have inherent flotation or thermal protection. Owners and captains sometimes view falls overboard and the failure to recover personnel as the result of crew member carelessness or inattention, overlooking nonuse of lifesaving equipment as a contributing factor.

Many fishermen consider their profession to be overregulated and are not receptive to requirements for costly equipment that does not contribute to their ability to catch fish (Correspondence on proposed rulemaking to U.S. Coast Guard, 1989; McCay et al., 1989). This attitude is particularly strong in fisheries with management practices that have significantly altered fishermen's fishing opportunities and negatively affected their income. Such perceptions were prevalent in comments from fishermen responding to the Coast Guard's Notice of Proposed Rulemaking, announcing prospective regulations for survival equipment prescribed by the Commercial Fishing Industry Vessel Safety Act of 1988 (CFIVSA). At the same time, the committee observed that fishermen routinely invest in items they consider important, such as the newest electronic navigation and fish-finding equipment (see Dewees and Hawkes, 1988; Levine and McCay, 1987).

Proficiency with Survival Equipment

Lifesaving equipment must be used quickly and correctly in order to be effective. It is not unusual for entire crews to be lost, even though their vessels are equipped with immersion suits and life rafts, or for immersion suits to be recovered *still in their carrying bags* or only partially deployed after an accident, suggesting someone unsuccessfully attempted to don the suit (see NTSB, 1989e). Such incidents occurred during this study (Alaska Fisherman's Journal, 1989b; Degener, 1989a,b):

- In November 1989, a 72-foot lobster boat equipped with seven immersion suits, a Coast Guard-approved life raft, VHF-FM and HF radios, a cellular telephone, and two emergency position-indicating radio beacons (EPIRBs) was reported overdue from offshore lobstering in Hudson Canyon (southeast of Long Island, New York). A signal from one of the vessel's EPIRBs was detected by a satellite and received by an Air Force ground station before it was reported overdue, starting an intensive air and surface search. Despite recovery of the EPIRB on the fourth day, multiple flare sightings, and sighting of what looked like a life raft below the surface, five fishermen and all other lifesaving equipment on board were not recovered (Booth, 1990; USCG, unpublished reports, 1989).
- In December 1989, a loaded clammer that had been in radio contact with the owner ashore disappeared in 6- to 8-foot seas during a snow squall near Cape May, New Jersey. A partially submerged, uninflated life raft and a partially deployed immersion suit were recovered. The vessel was located in 40 feet of water several days later. Three fishermen were lost (Degener, 1989a,b; Degener and Strawley, 1989).

Similar cases suggest that while many fishermen understand the potential benefits of lifesaving devices and have them aboard, many do not know how to properly use them (see Matsen, 1990; Nalder, 1990; Johnson, 1982). Familiarity with survival equipment was critical to recovery or loss of personnel during life-threatening situations. Some captains conduct abandon-ship drills and hands-on practice with safety and survival equipment (see Alaska Fisherman's Journal, 1989a), but the practice is not widespread (see Matsen, 1990; Nalder, 1990; Alaska Fisherman's Journal, 1988a,b; Miller, 1985).

Although many fishermen have had to use lifesaving devices, many others are not familiar with them. A study of marine safety in New Jersey found that almost half of the fishermen surveyed had never been in the water wearing a personal flotation device (PFD), only 34 percent had been in the water with an immersion suit on, and only 17 percent had ever inflated or used a life raft. Costs of repacking and the regional structure of the life raft inspection and repacking industry contributed to limited familiarity with life rafts (McCay et al., 1989). Although no similar research was found for other regions, the committee believes that the New Jersey findings generally reflect the state of familiarity with lifesaving equipment throughout the industry.

Survival equipment that was properly maintained, accessible, employed in time, and used properly by fishermen in good physical condition extended survival times sufficiently to permit rescue and recovery in almost all of the incidents reviewed (Alaska Fisherman's Journal, 1989a, 1988b; Zimmerman, 1989; NTSB, 1987, 1989c; Pollack, 1989; Johnson, 1982; Finley, 1982b; Matsen, 1990). Early deployment decisions provided the time necessary to overcome problems donning equipment and to conduct orderly abandon-ship procedures, including safe egress into life rafts, skiffs, or the water (McCay et al., 1989; NTSB, 1989a,b,c,d; Walter and Morani, 1986; Miller, 1985). Where any one of these factors was not present, the effectiveness of survival equipment was diminished or negated (Rostad, 1989b; Poggie and Pollnac, 1988; Castle, 1988; Alaska Fisherman's Journal, 1988b; Miller, 1985; Finley, 1982b; Johnson, 1982).

Despite the growing availability of survival equipment aboard many vessels beyond basic PFD requirements, anecdotal information indicates that it is often stowed out of sight or neglected until it is urgently needed (Nixon and Fairfield, 1986; see Matsen, 1990). This practice contributes to equipment failure. Some inferior equipment has been prone to failure (see Appendix I). Improper maintenance, unfamiliarity with use, and delays in employing the equipment played a significant role when available survival equipment did not perform to its advertised potential (NTSB, 1987, 1989a,b). Frequent problems affecting survival equipment use include inoperable zippers on immersion suits and improperly mounted or inaccessible life rafts (Matsen, 1990; Nixon and Fairfield, 1986; NTSB 1988b, 1989a,b,e; see Alaska Fisherman's Journal, 1989b). These problems are aggravated if fishermen have to learn how to use them during an emergency or do not have time to correctly deploy or correct problems with equipment before catastrophic loss of their vessel (Matsen, 1990; Rostad, 1989b; Alaska Fishermen's Journal, 1988b; NTSB, 1988b, 1989a,b,e; Nixon and Fairfield, 1986; Sullivan, 1984b; Johnson, 1982; Poggie and Pollnac, 1988).

Technical Issues

Safety and survival equipment required by federal regulations administered by the Department of Transportation (DOT) and its affiliated agencies must be "Coast Guard-approved," but not all equipment sold is approved (see NTSB, 1989a,c). How much nonapproved survival equipment is in use cannot be ascertained from existing data, but anecdotal references indicate that approved protective clothing is not widely used on deck (see Nixon and Fairfield, 1986). According to fishermen, the reason they do not buy or use Coast Guard-approved lifesaving devices such as antiexposure coveralls is that they are bulky, restrict their mobility on deck, or are not suited for wear in hot weather. In numerous falls overboard and sudden vessel losses, whatever the fisherman was wearing was his or her sole protection (Walker, 1990a; NCFVSI, 1989; Alaska Fisherman's Journal, 1988b; Rostad, 1989b; Zanoni, 1988; NTSB, 1989e, 1987; Dobravec, 1987; Nixon and Fairfield, 1986; Finley, 1982a; Gleason, 1982; Johnson, 1982; see Walker, 1990b).

Nonapproved Inflatable PFDs

Nonapproved inflatable equipment sold for recreational and commercial users is designed for convenience, such as color-coordinated jackets with yoke ("Mae West")-style air bladders hidden between the outer shell and inner liner. This equipment usually has a single air chamber, may not inflate to minimum Coast Guard standards for commercial use, may not have automatic activation devices to assist someone who is unconscious or disabled, and provides little protection against hypothermia. However, that they do not interfere with mobility is a strong selling point for deck work, and they are designed to be put on quickly in an emergency.

The Coast Guard is studying "in-service reliability" for continuous-wear inflatable PFDs. However, there is no market research by manufacturers that might indicate whether enough fishermen would purchase this gear so that it could be developed and sold at reasonable prices. Some nonapproved inflatable PFDs, worksuits, and immersion suits are designed with removable air bladders or flotation liners to facilitate cleaning. Outer shells can be worn without flotation liners attached. Devices of this type do not satisfy current Coast Guard standards.

Emergency Position-Indicating Radio Beacons

EPIRBs are electronic devices used to communicate distress to rescuers. A category 1, self-activating, 406-MHz satellite EPIRB incorporating updated technology was required on most uninspected fishing vessels operating on the high seas in May 1990 by federal regulations, with phase-in provisions. The conversion to the newer EPIRBs is motivated by the unreliability of the older classes of emergency beacons (Embler, undated; Lazarus, 1990b; Lemon, 1990b; The Westcoast Fisherman, 1989a; Pawlowski, 1987). The improved 406-MHz EPIRB is expected to greatly reduce the unresolved alert rate (98 percent of all alerts in 1989, which includes signals from aircraft, were unresolved [Coast Guard data]), providing sufficient signal reliability to justify immediate launching of rescue aircraft. The Coast Guard believes this will greatly improve rescue service to mariners in distress while concurrently reducing false alerts (Lemon, 1990b).

There are several issues of concern to fishermen. Category 1 EPIRBs are expensive, $1,700 or more. Category 2 EPIRBs have the same capabilities as category 1 EPIRBs but are manually activated and cost less. They may be carried as supplemental equipment, as may older style class B and C EPIRBs, which are still useful for homing purposes (see Commercial Fishing Industry Vessel Advisory Committee [CFIVAC], 1989; see NTSB, 1987, 1989a,b,c,d). They may be carried in lieu of a category 1 EPIRB in small, open vessels that do not have galleys or berthing facilities (Federal Register, 1990). Some fishermen who invested in the earlier models are irritated at having to purchase another EPIRB. They are also concerned that the costly new ones will be tempting targets for thieves (McCay et al., 1989). Similar concerns were expressed to the CFIVAC concerning life rafts. In New England, several high-quality, "free-floating" life rafts were stolen from their racks in 1989. Category 1, 2, and 3 satellite EPIRBs were intended to be registered with the National Oceanic and Atmospheric Administration (NOAA). Each comes with a postcard form for voluntary registration. Mandatory registration was not adopted by the Federal Communications Commission (FCC). Voluntary registration of EPIRBs and life raft serial numbers is being considered by the Coast Guard. Another issue is whether self-activating EPIRBs will work properly when used aboard small boats that are manufactured to stay afloat when swamped or capsized (C. Bond, personal communication, 1990).

However impressive this new technology, the end result may still be a search for bodies if the EPIRBs are not installed or maintained properly, or fishermen do not wear protective clothing or have access to a life raft to keep them alive until help arrives. Incremental implementation of safety initiatives, no matter how well intended, makes it difficult to ascertain their economic impact and overall contribution to improved safety. Even with conversion to the new EPIRBs, the absence of suitable protective deckwear leaves a significant gap with regard to the full range of equipment needed to enhance survival.

The Impact of Changing Technology

Fishermen, especially those with small-scale operations, often find it difficult to remain current on changes in technology. Many find it difficult to compare the relative value of equipment similar in appearance and function. Manufacturers and retailers confuse buyers by attaching Coast Guard-approved accessories to nonapproved protective clothing, failing to distinguish recreational and commercially approved devices in their promotional materials, including lifesaving devices approved only for recreational use in industrial catalogs, selling survival equipment not approved by the U.S. Coast Guard, and making misleading references to Coast Guard approval status (e.g., "not USCG-categorized").

Promotional literature sometimes encourages the use of nonapproved in lieu of approved equipment. For example, one manufacturer's promotional literature attributes the following quote to a Coast Guard technical authority: "If these garments work for your application, then you should seriously consider them. Protection you wear is much better than approved devices you keep under a seat or a locker."

Except for special requirements for commercial hybrid PFDs and immersion suits, and pamphlets for dual recreational-commercial-use PFDs, pamphlets, manuals, videotapes, or other instruction aids for safety and survival equipment have not been required or approved by the Coast Guard for use aboard uninspected commercial vessels. Except for basic information of varying thoroughness about products, such materials are not usually provided by manufacturers with this equipment. Thus, some fishermen don't have a sufficient frame of reference for determining what is good, what is not, what will satisfy their safety needs within their budgets, and what will satisfy existing and prospective Coast Guard requirements (see Alaska Fisherman's Journal, 1989a; NTSB, 1987).

Fishermen who have participated in safety or survival training programs using a variety of gear appear better prepared to assess the relative value of lifesaving equipment. Even then, some of them choose not to purchase or use certain equipment. Some fishermen are delaying purchasing decisions until the Coast Guard issues its final regulations under the CFIVSA. Others cite infrequent encounters with hazardous conditions, inshore operating areas, costs, and limited onboard storage space (see McCay et al., 1989; NTSB, 1988a).

The underlying reasons that fishermen don't carry the latest lifesaving gear are complex. Research in this area is limited, but suggests that technology adoption is related to economic advantages, perceived risk, tradition, long-established or inflexible work habits, and simplicity of use. The research further suggests that varied individual attitudes and perceptions, technological advances, innovations, fisheries, and economic circumstances require a situational approach to implementing strategies requiring adoption of new technology (Levine and McCay, 1987; Dewees and Hawkes, 1988).

Gaps in Survival Equipment

Protective Deckwear

Over 61 percent of all fatalities recorded by the Coast Guard between 1982 and 1987 were fishermen who fell into the water or disappeared for unknown reasons. Most of them probably were not wearing lifesaving equipment or protective clothing, or their vessels were suddenly overwhelmed by catastrophic events. The reasons for not wearing existing equipment have some merit. The inherent flotation in most work suits is bulky, and use of work suits containing flotation devices is impractical for the manual work conducted aboard small vessels. Fishermen routinely use rainwear that is practical and convenient, but practical protective clothing that is also convenient to wear and offers full-time lifesaving protection is not widely available on the U.S. market. Some gear that is available (such as inflatable flotation coats) has not been submitted to the Coast Guard for approval.

The CCG's fishing vessel safety study found that a special work suit for fishermen was needed in lieu of the more traditional "keyhole"-type life jackets (CCG, 1987). Canadian standards would improve thermal protection while maintaining flotation capability, without impairing mobility during normal work (Canadian General Standards Board, 1989). The new standard is specifically intended to result in an attractive and affordable alternative to commonly worn rainwear. Regulations requiring work suits meeting the new standard were scheduled to be in place by the end of 1990. The Canadian standard permits either a one- or two-piece system; a two-piece work suit meeting the standard is currently being marketed (promotional materials). The U.S. Coast Guard has an approved standard for two-piece wet suits for special applications. This provides a precedent for considering approval of a two-piece fisherman's work suit should manufacturers choose to seek formal Coast Guard approval.

Survival Platforms

U.S. fishermen have expressed considerable concern over the need to equip vessels operating inshore with life rafts, especially where they typically operate in sight of land. However, SAR data demonstrate that all fishermen, including those working inshore and on inland waters in coastal regions, are exposed to life-threatening situations to a larger degree than many acknowledge. Storage space aboard a small vessel can be extremely limited, so highly portable life rafts or survival platforms suitable for small fishing vessels merit consideration. The U.S. Coast Guard is developing proposed rules in this regard.

As part of the Canadian mandate to improve fishing vessel safety, the CCG Ship Safety Office is also overseeing development of a prototype coastal life raft especially designed for small fishing vessels. The prototype is still in early stages (see Appendix I). It would be prudent to monitor the results of this project, as well as experience with the new fishermen's work suit system, for potential application aboard U.S. fishing vessels.

STRATEGIES FOR ADDRESSING SURVIVAL PROBLEMS

Major survival problems uncovered during this study include:

- inadequate protective clothing suitable for routine use on the deck of a fishing vessel,
- inadequate or insufficient survival equipment,
- inadequate routine use of protective clothing,
- weaknesses in knowledge and practical skills in maintaining and using survival equipment, and
- weaknesses in timely decision making to employ survival equipment.

Most survival problems can be addressed as either a personnel- or vessel-related issue, with the exception of suitable protective clothing for use on the deck of a fishing vessel. Correcting this problem could be approached as a research and development project leading to suitable standards and prototype equipment and subsequent marketing of a commercial product.

The committee has identified three survival-specific safety-improvement alternatives (the sequential numbering begun in Chapter 3 continues) as follows.

Alternative 23: Require Manufacturers to Provide Installation, Maintenance, and Use Instructions

This alternative would expand the Coast Guard's existing requirements for instructional materials for survival equipment to include materials specifically designed for fishermen. Both written materials and, in most instances, instructional videotapes could be required. They would provide basic technical and performance information, maintenance instructions, and installation, activation, donning, and other instructions. Although not a substitute for practical training, videotapes could provide a way to convey important information about survival equipment, including a demonstration of actual use.

Alternative 24: Develop and Require Carriage of Fishing-Industry-Specific Survival Equipment

This alternative envisions an approach like that taken by the Canadian government, which has developed a standard for a work suit for the fishing industry, has worked with the manufacturers of lifesaving equipment to bring the suit to market, and is establishing regulations requiring these work suits aboard each commercial fishing vessel in sufficient number for all aboard. Fishermen, as a rule, do not wear protective clothing with flotation or thermal protection, or clothing, including headgear, designed to guard against occupational hazards. The usual complaint is that the available protective clothing is difficult, inconvenient, or impractical to wear while working on a fishing vessel. It is also true that some fishermen do not wear protective clothing regardless of how well designed and constructed it is. There is an urgent need for protective, convenient-to-wear gear.

In this alternative, the U.S. Coast Guard could develop parallel standards for a work suit, inflatable PFDs, and similar equipment and, if necessary, fund prototype development to bring this equipment to market. The Occupational Safety and Health Administration (OSHA) and the Coast Guard, in consultation with the commercial fishing industry, could jointly develop personal workplace safety equipment suited for the fishing industry. Implementation issues include accepting prototype development as a Coast Guard or OSHA responsibility, and funding.

Alternative 25: Prohibit Use of Survival Equipment That Is Not Coast Guard-Approved

This alternative envisions that only equipment approved by the Coast Guard would be permitted for carriage aboard commercial fishing vessels. The logic is that because this equipment must meet strict standards, the probability that it will work correctly on demand would be greater than that for nonapproved equipment, for which there is no Coast Guard oversight during development and manufacture. This alternative, if adopted, would necessarily have to follow development of a full range of suitable survival equipment convenient for use aboard commercial fishing vessels. This would be necessary so that levels of protection available through certain conveniently used equipment not bearing Coast Guard approval, yet used by some fishermen, will not be lost. The fishing industry as a market segment may not be sufficient by itself to support this alternative. Prior to implementation, the benefits and costs would have to be carefully analyzed to ascertain the effects on development and marketing of survival equipment.

SUMMARY

The problems associated with survival equipment are that the equipment is not available, not used at all, not used in time, or not used properly or fails to perform as intended. Nonuse of PFDs was frequently associated with man-overboard fatalities. Fishermen throughout the industry do not ordinarily wear lifesaving devices of any type while working on deck, either during transits to and from the fishing grounds or while fishing. Fishermen operating in environmental conditions where exposure is a major factor usually wear some type of protective clothing. However, protective clothing with thermal or inflatable protection that is designed for convenient wear in the fishermen's workplace has been developed and marketed in Canada. This equipment, as of August 1990, has not been submitted for U.S. Coast Guard approval.

Other nonapproved safety and survival equipment designed for special purposes or convenience of the wearer is available. Equipment not approved by the Coast Guard may be carried aboard uninspected vessels as optional equipment. The quality of nonapproved safety and survival equipment is highly variable. The effectiveness of this equipment in survival situations is not known.

Safety improvement alternatives (continued from preceding chapters) that could be employed are:

23. require manufacturers to provide installation, maintenance, and use instructions,

24. develop and require carriage of fishing-industry-specific survival equipment, and

25. prohibit use of survival equipment that is not Coast Guard-approved.

7

External Influences on Safety

The committee considered three external influences on safety: fisheries management practices, insurance, and environmental conditions. Although neither fisheries management practices nor insurance directly causes vessel or personnel casualties, each contributes to an economic environment that has a potential impact on fishing vessel safety.

FISHERIES MANAGEMENT INFLUENCES ON FISHING INDUSTRY SAFETY

Fishing has been romanticized as the last frontier, a natural environment where fishermen see themselves as hunters who work independently and have life-styles outside of traditional occupational patterns. The implied freedom from outside interference and influences is a myth. While the work takes place in relative isolation, harvesting of most fisheries is highly regulated (Sutinen et al., 1989a,b; Chandler, 1988; Mollet, 1986; Nies, 1986; National Oceanic and Atmospheric Administration [NOAA], 1986; Pontecorvo, 1986; Frady, 1985; Grasselli and O'Hara, 1983; Maiolo and Orbach, 1982; Browning, 1980).

Fishery regulations are enforced through vessel-boarding programs conducted by the National Marine Fisheries Service (NMFS) and the U.S. Coast Guard, and in cooperation with state authorities (Sutinen et al., 1989a,b; Chandler, 1988; NOAA, 1986; Nies, 1986; Nies and Carney, 1985; Sutinen and Hennessey, 1986; Grasselli and O'Hara, 1983). The eight regional fisheries management councils (FMCs, described in Chapter 2) have had difficulty accommodating effective conservation and conflicts over fishery allocations and

Coast Guard boarding officer checking mesh size. (PA2 Robin Ressler, *U.S. Coast Guard*)

safety. The management system's inability to match harvesting capacity to biological productivity of fishery stocks has resulted in a highly competitive operating environment in which fishermen may take unnecessary risks to maintain their livelihood. This practice has resulted in overcapitalization in some fisheries and more marginal operators who find it economically difficult to adequately maintain and equip their vessels to improve safety in a hostile environment (Wiese, 1990; Chandler, 1988; Pontecorvo, 1986; Lassen and Van Olst, 1986; Couper, 1985). Furthermore, early closures, short seasons, and selective gear allocations have caused many operators to abandon the fisheries their vessels were designed for and enter fisheries far from home port, change to new fishing gear, or enter entirely new fisheries. Under these circumstances the vessel operator and crew may face new risks and potential safety problems.

Under the regional council system established by the Magnuson Fisheries Conservation and Management Act (MFCMA), comprehensive fisheries management plans (FMPs) are prepared using the concept of optimum yield (OY) as the standard. OY is defined as maximum sustainable yield (MSY) modified by economic, social, and political factors. In virtually all cases in which OY was determined to achieve sustainable yield, the target MSY has been subsequently raised to accommodate fishermen's needs. This practice increases pressure on fish stocks and results in short-term overfishing and decline in abundance (see Gordon, 1989; Siegel and Gordon, 1989; Chandler, 1988; NOAA, 1986). There have been few effort-limitation schemes to balance harvesting capacity to biological production in the United States, and they have been narrowly applied (see Mollet, 1986).

The management councils are also charged by the MFCMA to consider safety in their FMPs. Section 303(a)(6) of the MFCMA as amended states [that the FMCs] "consider and may provide for temporary adjustments after consultation with the Coast Guard and persons utilizing the fishery regarding access to the fishery for vessels otherwise prevented from harvesting because of weather or other ocean conditions affecting the safety of vessels." The FMCs clearly have the opportunity to address safety factors related to weather or ocean conditions within each management plan. Yet, in virtually every case, safety has been subordinated in favor of economic interests. Where conservation decisions and use determinations are combined, constituents exert pressure to serve more users (see Chandler, 1988; NOAA, 1986; Lassen and Van Olst, 1986). Competition intensifies as returns decrease, fishing time is reduced to achieve quotas, and more fishermen compete for fewer fish. With stiffer competition for dwindling resources, more fishermen take risks and accidents increase. Similar occurrences are associated with certain internationally managed fisheries, involving in-season, time, and area openings and closures. For example, short, inflexible openings associated with halibut and salmon in the West Coast and Alaska regions (i.e., "derby-" or "olympic"-style fishing) have at times effectively forced fishermen to work under extremely adverse environmental conditions or not at all. Vessels and lives have been lost as a result (see Parker, 1990; Alaska Fisherman's Journal 1987; Fishermen's News, 1988; Sitka Sentinel, 1987; Lassen and Van Olst, 1986). The data and anecdotal information are inconclusive about whether the number of incidents for olympic-style fishing is significantly higher than might have occurred during an extended season.

The multitude of accidents in some fisheries has prompted reconsideration of safety. Some regional councils attempted to improve safety by revising management practices; for example, by adding fishing days to accommodate bad weather, such as in the Atlantic surf clam fishery. The rationale is that additional fishing days contribute to safety if vessels limit their exposure to adverse weather. However, there are no data to evaluate the effectiveness of this practice (see Lassen and Van Olst, 1986).

Other management innovations to improve economic conditions have been or are being considered; for example, limited-entry programs that restrict the number of licenses and individual transferable quota (ITQ) systems that redefine targeted fisheries as quasi-private property and award percentage shares of annual fishery quotas to individuals or firms. Generally, economic measures have been approached on a fishery-by-fishery basis (see Mollet, 1986; Frady, 1985). Measures that do not restrict vessel conversions could create or aggravate redundant harvesting capacity in other overcapitalized fisheries, potentially shifting safety problems from one fishery to another. Furthermore, there are no guarantees that vessel owners who experience improved economic returns as a result of fishery management innovations will invest in safety training or equipment or in improving the physical condition of their vessel, nor is there any federal requirement to do so.

A full examination of these concepts and other fisheries management practices and related economic factors that contribute to accidents is beyond the scope of this study, but the three alternatives that follow (continuing the sequential numbering begun in Chapter 3) could provide improvement in safety opportunities.

Alternative 26: Establish Flexible Season Openings

This alternative builds on the existing movement in some regions to incorporate flexibility in season openings to provide alternative fishing days when marginal or adverse weather is forecast or occurs during openings. Implementation of this alternative could make managing those fisheries more complex, however, and might increase infrastructure requirements, particularly monitoring to support decision making. The alternative's benefits are that once the implementing authority and arrangements are in place, it could be activated quickly. It is not a widespread solution, as it applies to a specific issue in only a few fisheries that are seasonal in nature, such as Pacific salmon.

Alternative 27: Establish a Voting Position for a Marine Safety Organization on Each Fishery Management Council

Safety issues are represented in regional councils by a nonvoting Coast Guard representative, typically a staff officer below flag rank. Providing a voting member of high seniority from a marine safety organization on each of the eight regional councils could upgrade attention to safety as an important element of fishery management. This alternative envisions making a Coast Guard flag officer a voting member on each of the councils for the express purpose of addressing safety issues.

Alternative 28: Expand Safety Emphasis in Fishery Management Plans

Although the MFCMA was amended to provide for safety considerations because of weather or oceanic conditions, councils in general have not fully accommodated safety in recommending management regimes to the Secretary of Commerce. As more and more fisheries approach maturity, fishermen are constrained by conservation provisions of the management plans, and economic pressures on them mount. Faced with escalating costs and a near-stable or declining resource base, they are frequently forced to minimize maintenance, which has implications for safety. Councils tend to consider management on a fishery-by-fishery basis and do not look at the broader aspects of total fishery management regimes.

This alternative envisions that safety be upgraded to National Standards status within the MFCMA to require greater attention to safety issues. It further envisions that this could lead to managing fisheries in ways that would create economic opportunities that support safety improvement as a secondary objective. Achieving secondary objectives through programs whose principal objectives lie elsewhere is not uncommon in federal policy. The data provide an insufficient basis to support fishery management regimes for safety purposes alone, but as FMCs are charged to consider safety aspects, greater effort could be made to evaluate fisheries for such considerations.

INSURANCE

Fishing vessel safety and insurance have become inextricably linked. When fishermen are distressed that they cannot buy insurance at affordable prices, they are acknowledging that their collective safety records have reached intolerable levels, because the losses that comprise those records determine insurance costs. It must be noted that insurance does not reduce or eliminate losses; it reduces only the financial risk associated with them. An insurance buyer purchases protection from risk by paying a premium—in essence, his share of the redistribution of losses over all those who have sought similar protection. Thus, risks are mitigated, but losses are not. They remain an economic drain on society, and the premiums that reflect this aggregate drain rise or fall with the losses.

Some form of insurance is needed by commercial fishing vessel owners to protect themselves against loss or damage to their vessels and potential financial liabilities that can result from injuries or damage to others, including their own crew members. Coverage may be obtained from commercial insurers or through membership in mutual associations, exchanges, or "clubs" formed for group self-protection (Nixon, 1986; Nixon et al., 1987).

The aggregate premium paid for any one class of business (e.g., fishing vessels) should reflect the cumulative losses within that class. Assureds with records of low claim frequency and severity may be rated more favorably than those whose records reflect higher losses, but all must contribute toward the funding of major casualties, that is, those of such magnitude that they cannot be borne by a single assured. Insurance premiums take into account not only this redistribution of losses but also administrative overhead expenses, the profit element (whether from underwriting results or investment returns), and competition in the insurance market. It is against this backdrop that the problems associated with fishing vessel insurance emerge.

Availability and Affordability of Insurance

During the past three decades, concerns about insurance availability and cost to fishing vessel owners led to government studies (Danforth and Theodore, 1957; Lyon and Theodore, 1976) and to the formation in 1978 of the National Council of Fishing Vessel Safety and Insurance (NCFVSI). The council was established by representatives of the fishing industry, marine insurers, and others from universities and the private sector. Its goals are to improve the safety of vessels and personnel and to reduce insurance costs to the fishing industry (Lassen, 1985).

Beginning in 1984, joint concerns about commercial fishing vessel safety records and the cost and availability of liability insurance led to enactment of the Commercial Fishing Industry Vessel Safety Act of 1988 (CFIVSA, P.L. 100-424) and this study. From the fishing industry's standpoint, the primary concern has centered around availability of insurance at affordable prices (Nixon, 1985, 1986; Pacific Fisheries Consultants, 1987). The fishing industry as a whole has presented underwriters with marginal to unsatisfactory losses. However, many people in the fishing industry believe that insurance companies have not paid adequate attention to differences in quality among fisheries, fleets, and owners in establishing equitable rates. Moreover, fishermen believe that the insurance industry has not provided stable, dependable rates and availability of insurance, particularly during the 1980s (Nixon et al., 1987; Nixon, 1986; Sullivan, 1984a,d; Clendinen, 1984).

From the underwriter's perspective, early 1980s market forces disrupted competition in the insurance industry and exacerbated already deteriorating market conditions. Marginal underwriting results occurred before 1980, yet competition kept fishing industry insurance rates depressed. Serious underwriting losses followed. The American Institute of Marine Underwriters (AIMU), whose 120 insurance company members underwrite about 90 percent of marine insurance in the United States, estimated that fishing vessel insurance represented only about 5 percent of its members' marine premiums. AIMU observed that many companies writing fishing vessel business were not among its members. Nevertheless, as the only national marine insurers' organization in the country, it undertook a survey of members writing fishing vessel insurance. It found that between 1980 and 1984, underwriting losses and expenses amounted to 156 percent of premiums received (against an overall marine insurance loss ratio of 107 percent). Moreover, the fishing vessel loss ratio was never less than 138 percent in any one year, evenly spread among regions (AIMU, 1985). This imbalance was intolerable to some insurers, and a number withdrew from the market. Those who remained concluded that rate correction could not be postponed. The result was a smaller, more select, higher-priced insurance market (Pacific Fisheries Consultants, 1987; AIMU, 1985; Glickman, 1984; Sullivan, 1984a,d; Mohl, 1984).

Over the years, in the face of similar cycles, some fishing vessel owners banded together to establish mutual associations for self-protection. Several such mutuals were formed, usually with support from commercial reinsurers to protect against large, catastrophic losses that would financially strain the group. Most important to their long-term success is the ability to find and maintain members whose vessels and operations present similar risks and who have similar attitudes on safety and loss prevention. Like-minded members possess a peer review capability that can generate strong incentives for enhanced risk-management activity, including adopting and using common operating and maintenance standards and inspection requirements (Melteff, 1988; Nixon et al., 1987; Nixon, 1986).

Good safety performance results for hull and machinery insurance are reported for several self-insurance pools. These pools' strict technical and outfitting requirements (Hamilton, 1989) sometimes exceed Coast Guard requirements (National Transportation Safety Board [NTSB], 1987; Pacific Fisheries Consultants, 1987; Nixon et al., 1987). Equipment aboard vessels in such pools has contributed to survival in major accidents like fires and sinkings (Yoder, 1990).

Despite the best prevention efforts, however, accidents will occur. Resultant injuries and fatalities involve medical costs, which have increased steadily over time. For many years, medical costs to commercial fishermen and other mariners aboard U.S.-documented vessels were subsidized by the federal government. The U.S. Public Health Service (USPHS) conducted physical examinations, issued fit-for-duty certificates (often required by fishing vessel owners prior to taking an individual aboard as a crew member), and treated injuries at no direct cost to the fishing industry (Wiese, 1988).

The elimination of USPHS medical facilities and services beginning in 1981 removed the opportunity in some ports for physical screening after hospitalization or treatment and aggravated the impact of medical costs borne by the private sector (commercial insurers, self-insurance pools, and individual vessel owners). Moreover, liability exposures, particularly for pain and suffering, continue to threaten both commercial underwriters and self-insurance pools because of the potential size and unpredictability of jury awards. A significant portion of liability claims are associated with federal court cases where litigants can take advantage of both the doctrine of unseaworthiness, which utilizes strict liability standards, and the Jones Act—even for short-term temporary disabilities (Nixon, 1986, 1985; Nixon and Fairfield, 1986; AIMU, 1985).

Loss-Prevention Leadership

In regional assessments, some fishing industry representatives expressed disappointment that underwriters have not assumed a national leadership role in establishing standards for "safe" vessels. Underwriters of fishing vessels believe that government agencies and the fishing industry itself are better poised to assume such leadership. Their reasoning is that if they were to assume the initial costs in setting and enforcing standards, they would have to pass them along to their customers. A practical concern is that this would be difficult in the recurrent competitive atmosphere of international insurance markets.

Marine property and casualty business today generates less than 3 percent of premiums written by U.S. property insurance companies (Best's, 1988). This segment of the business is having difficulty maintaining its identity as a distinct class of business in some companies (AIMU, 1988). Moreover, the market is fragmented. As noted earlier, members of the only national marine insurers' organization underwrite only a small fraction of the country's fishing vessel business. In a competitive climate, what one underwriter may require as a standard, another may not. Companies who are now insuring the fishing industry tend to favor leaving decisions to local underwriters, who know their customers and the risks and improvement opportunities applicable to their locale. In summary, the underwriting community does not consider itself in a position to exert cohesive, national leadership in the field of fishing vessel safety.

Unresolved Problems Affecting Underwriting

Underlying problems associated with fishing vessel insurance can be grouped into three areas: the safety record, the data bases, and the tort system. Regarding the safety record, underwriters acknowledge the high risks inherent in fishing, but have expressed the following specific concerns (AIMU, 1985):

- the so-called "moral hazard," a concern that the vessel will be sunk, stolen, or burned in order to claim the insured value (such losses have occurred and other attempts have been alleged; Lazarus, 1990a; Providence Journal, 1990; Letz, 1986; see Salit, 1989; Clendinen, 1984; Sullivan 1984c,d);
- risks of fishing in adverse weather and seas;
- vessel stability;
- utilization of vessels in fisheries other than originally intended;
- inadequate maintenance of vessels and equipment;
- lack of design, construction, or safety feature standards;
- lack of inspection requirements other than those imposed by the underwriters, who do not see themselves as enforcers of standards or inspection requirements; and
- lack of licensing requirements or other evidence of professional competency.

The lack of uniform statistics and reliable data bases is another major problem in analyzing claims. However, the NCFVSI, in conjunction with the Marine Index Bureau and volunteering underwriters, has begun a comprehensive data base of vessel casualty and personal injury statistics, known as the Commercial

Fishing Claims Registry (CFCR) (see Appendix D). It is in a formative stage and of very limited utility for this study. However, the CFIVSA mandates a system to compile marine casualty statistics from underwriters' data, so a nationwide uniform data base appears to be on its way toward implementation. Whether similar meaningful data will be obtained from uninsured and self-insured interests or from underwriters internationally is uncertain, but any data base will be seriously deficient without that input. Further, the data base output capability needs to be carefully structured to be most useful. Recognizing that individual input sources may need to remain confidential, output that will permit regional, local, and even fishery-specific analyses could be most beneficial to underwriters.

It is not within the scope of this assessment to make recommendations concerning the U.S. tort system or legal remedies for maritime injuries. Indeed, Congress has considered this subject on many occasions, most recently in hearings on the CFIVSA. What cannot be ignored is that the current legal framework for compensating seamen's injury claims has had a serious impact on the capability of insurers to establish rates and provide liability coverage for the fishing industry. As previously mentioned, the size and unpredictability of awards are among concerns expressed by underwriters. No change is likely to occur in the legal system in the foreseeable future. Thus, fishermen who must insure against potential liability should be motivated by that realization alone to do whatever they can to reduce the likelihood of accidents.

Improving Communications

Reductions in insurance costs can only follow reductions in vessel losses and crew claims. Recognizing these economic considerations should be a major motivation for vessel safety. There is evidence in the industry—particularly in self-insurance pools—that this correlation has occurred and that fishermen are capable of taking corrective measures, at least as far as vessel hull and machinery claims are concerned. Spreading this word throughout the fishing industry is logical and beneficial (alternative 14). The NCFVSI has already established this goal, but expanding the effort through education and training programs (alternative 19) has merit as well. One new alternative directly related to insurance remains to be discussed.

Alternative 29: Mandate Compulsory Insurance

During this study, the idea of compulsory insurance was suggested as a way to motivate attention to the material condition of vessels and equipment. The logic is that to obtain insurance, owners would have to upgrade the condition of their vessels to acceptable standards. There are insufficient data to evaluate this alternative; however, there are several major implementation issues. As discussed in this chapter, the insurance industry does not regard itself as having, or being in a position to assume, a central role in improving safety in the fishing industry. Additionally, as a practical matter, the structure and condition of the insurance industry are such that implementing this alternative could be very difficult.

MARINE WEATHER SERVICES

A storm at sea is an awesome occurrence, and adverse weather has been an element in many fishing vessel casualties (see Waage, 1990; Zimmerman, 1989; Fishermen's News, 1988; Alaska Fisherman's Journal, 1988b; NTSB, 1987; Sitka Sentinel, 1987; Ball, 1978). Yet, weather was not implicated as a major cause of fishing vessel accidents in the Coast Guard main casualty (CASMAIN) and search and rescue (SAR) data. The majority of SAR incidents, over 66 percent, occurred when winds were under 20 knots and seas not greater than 3 feet; 88 percent occurred when winds were less than 20 knots and seas not greater than 10 feet. SAR data list environmental conditions as the principal cause of under 4 percent of the cases. CASMAIN attributes less than 4 percent of all commercial fishing casualties to weather and sea conditions. One reason the data do not implicate weather more strongly is that they focus on primary rather than ancillary causes. Weather as an element in accidents is poorly tracked, but anecdotal information suggests that while it may not have been the proximate cause, it was clearly an important contributing factor in many casualties.

Information on how weather forecasting influences decision making by fishermen is scarce. However, a recent National Research Council (NRC) study found that fishermen generally view forecast services as adequate (NRC, 1989a). Fishermen supplement this information by exchanging weather observations with other vessels, and they rely heavily on their own instincts and interpretations, which vary with experience and local knowledge. For example, vessel captains operating in the Aleutian Islands, Alaska, or transiting Puget Sound, Washington, need to know about currents and wind speed and direction, which can create effects like severe surface turbulence. On the Atlantic and Pacific coasts, operators must be cognizant of weather conditions that make bar crossings extremely treacherous. Reliable weather information is thus particularly important when a storm system is approaching and conditions are changing rapidly.

The value of local weather information is vividly demonstrated by services provided in the Kodiak Island and Dutch Harbor, Alaska, areas initiated voluntarily in the mid-1970s by a Kodiak fisherman's wife, Mrs. Peggy Dyson, in cooperation with the National Weather Service (NWS). Fishermen credit her locally revised forecasts with providing them information in time to seek shelter to prevent major damage or loss of their vessels during rapidly changing weather conditions. Mrs. Dyson has provided a vital service to Coast Guard rescue forces by interpreting weather data to approximate locations and on-scene conditions in potentially life-threatening situations (Rostad, 1989a). Local weather broadcasts are available in varying degrees in other locations as well, including fishing ports in New England, the mid-Atlantic, the Gulf Coast, and northern California and Oregon (Westcoast Fisherman, 1989b).

Opportunities to improve marine weather services for the fishing industry were identified in a 1989 NRC report (NRC, 1989a). The committee's collective experience is that environmental conditions are an important contributing factor in fishing vessel casualties and that up-to-date weather information could be used by fishermen to improve their margins of safety. An alternative for improving safety is providing local weather services specifically designed to support the fishing industry.

Alternative 30: Expand Fishing-Industry-Specific Weather Services

The thrust of this alternative is to extend fishery advisory services containing weather information to cover fishing grounds and ports where such services might prove useful. Infrastructure needs include access to the NWS and pertinent weather data, a marine radio station, and funding. Services of this type could be provided by volunteers, the fishing industry, the government, or a combination thereof.

SUMMARY

Fisheries management practices and environmental conditions contribute to fishing vessel casualties, but the nature and scope of each as a causative factor are unclear. Local fishing-industry-specific weather services have been effective in improving the safety of fishing vessel operations.

The availability and cost of insurance on fishing vessels are directly related to high losses. The fragmented insurance market for this class of business in a competitive economic climate has had difficulty establishing and maintaining a realistic rate base, partly because of lack of claims statistics and the uncertainty of liability awards. Increased interindustry communications, a better data base, and measures to reduce fishing industry losses are the most meaningful safety improvement options to employ.

The alternatives (continued from preceding chapters) offered are:

26. establish flexible season openings,
27. establish a voting position for a marine safety organization on each fishery management council,
28. expand safety emphasis in fishery management plans,
29. mandate compulsory insurance, and
30. expand fishing-industry-specific weather services.

8

Conclusions and Recommendations

Commercial fishing, inherently a dangerous undertaking, has one of the highest mortality rates of any occupation. Furthermore, a substantial number of vessels are lost and many more are damaged or break down each year during fishing operations and transit to and from the fishing grounds. Overall, the industry's safety performance record is so poor that the availability and cost of insurance have become major sources of concern to many fishermen. Despite these facts—unlike most other maritime activities—the safety of fishing industry vessels has, until recently, gone largely unregulated. Voluntary measures relied on to improve safety have been spotty and inconsistent, though if universally applied, some appear to have significant potential to improve safety performance.

The fishing industry's safety record can be improved, but this will require mandated, systematic attention to safety throughout the industry. Greater federal involvement will be required to bring all safety measures used into a cohesive and effective program. Near-term implementation of basic safety measures is feasible by modestly expanding or building on existing resources and coordinating efforts by federal and state governments, fishery management councils, fisheries commissions, industry, and interested third parties in the administration process. Anticipated benefits for all elements of the safety program need to be balanced against the costs. In some cases, congressional authority will be required to enable implementation.

This study has identified safety problems and issues in five general areas:

1. safety administration,
2. vessel fitness,

3. human factors,
4. safety equipment and survival, and
5. external influences.

Conclusions and recommendations in each of these areas respond to the congressional intent under the Commercial Fishing Industry Vessel Safety Act of 1988 (CFIVSA, P.L. 100-424) to reduce the incidence of casualties, fatalities, and injuries.

Unfortunately, the degree to which the safety measures recommended in this chapter will reduce vessel and personnel casualties cannot be estimated precisely. While the apparent rate of incidence is high, the industry is small, and the aggregate number of vessel- or life-threatening events is low relative to that of other high-risk industries. One or two large-scale fishing industry vessel disasters can skew casualty rates in any given year. Moreover, it is difficult to know how many events would occur regardless of safety precautions—just as it is impossible to know how many events do not occur because safety preparations succeed. Dramatic reductions in casualties may not result immediately, and decreasing casualty trends may prove difficult to discern in the short term. However, meaningful and measurable improvements can nevertheless eventually be achieved.

ESTABLISHING AN INTEGRATED SAFETY PROGRAM

Safety problems have generally been approached individually rather than systematically, resulting in what at best is a partial solution. A holistic approach is needed to ensure that the full nature of each problem is considered, an appropriate range of alternatives for addressing the problem is developed, and balance is maintained with other elements of an overall safety-improvement strategy. A "total concept" integrated program would include:

- goals and objectives set to achieve improvements in safety nationwide, but refined to take into account regional variations in exposure to hazards, operating and working conditions, safety performance, and other relevant factors;
- a data base sufficient to identify problems, evaluate improvement alternatives, and monitor results;
- standards of performance, both of vessels and of personnel, established with the aim of meeting safety objectives while at the same time improving the quality and potential productivity of the nation's fishing fleets;
- means to achieve and maintain these standards, e.g., training programs and equipment research and development;
- means to monitor and enforce the standards and regulations, e.g., operator licensing and vessel inspection; and
- a deliberate methodology to evaluate program effectiveness and progressively introduce adjustments as needed.

SAFETY LEADERSHIP

Industry and individual leadership have certainly contributed to safety within the industry, but have not been able to sufficiently motivate universal attention to the subject. Voluntary private and government initiatives have been highly fragmented and have induced only small response. Strong, central leadership underwritten with sufficient resources to sustain long-term momentum is needed to provide an effective forum for addressing safety on an ongoing basis and to implement and administer a comprehensive, integrated safety program.

RECOMMENDATION
Establish Federal Leadership

The Department of Transportation, acting through the Coast Guard, should lead a coordinated national effort to improve safety within the commercial fishing industry. The National Oceanic and Atmospheric Administration (NOAA), Occupational Safety and Health Administration (OSHA), international and national fisheries commissions, states, the fishing and insurance industries, marine educators, and other interested or affected parties should, within their areas of responsibility or service, cooperate fully with the Coast Guard in establishing the national, regional, and local leadership and resources necessary to improve safety in the fishing industry.

IMPLEMENTING THE PROGRAM

Significant levels of manpower and financial resources would be required to implement the more stringent safety alternatives, such as vessel inspection. The prospective effectiveness of these more costly alternatives cannot be assessed within the scope of existing data. Less-resource-intensive safety interventions, grounded in existing programs, could facilitate near-term implementation and have potential to increase safety awareness and improve vessel fitness for service. At the same time, they would lay a foundation for more-stringent safety measures in the longer term if it becomes clear that they are necessary to meet safety objectives. As a cautionary note, higher casualty rates may be disclosed as more complete data are developed. Therefore, the effects of data-improvement regimes must be considered when evaluating progress in meeting safety objectives and determining whether to impose more rigorous safety-improvement alternatives.

RECOMMENDATION
Implement an Integrated Safety Strategy by Stages

The Coast Guard should implement a comprehensive safety program that addresses, in stages, the full range of safety problems. Initial program elements should impose the least onerous burden on the fishing industry—insofar as possible—maximizing use of relatively low cost, least intrusive measures that can be implemented quickly using existing resources. The effectiveness of the measures taken should be evaluated as data are developed. If unsatisfactory or ineffective for some or all categories and sizes of vessels, more-stringent measures should be considered and introduced in stages where needed until desired safety-performance objectives are achieved.

SAFETY ADMINISTRATION

Effective safety administration will require a comprehensive program that first identifies and encourages participation of all who have the potential to contribute to the program, evaluates this potential, and organizes a safety infrastructure such that all efforts are mutually supportive and integrated to achieve maximum results. Next, means must be found to acquire sufficient and usable data to identify specific safety problems, measure their impact, evaluate improvement alternatives, and monitor results.

In this regard, the data are incomplete concerning vessel employment in commercial fishing, the population at risk, exposure levels (including changes resulting from advances in technology), the full scope of vessel and personnel casualties and accidents, and personal injuries. These and other factors limit the utility of comparative analysis with other occupations. The data are sufficient to conclude that opportunities exist to improve safety, but do not provide sufficient insight on cause-and-effect relationships to predict the effect of alternatives to achieve safety-improvement objectives. Therefore, implementation of safety alternatives needs to be coordinated with development of means to evaluate the results.

RECOMMENDATION
Upgrade Safety Administration

The Coast Guard should upgrade the capability to administer an integrated safety program. The Coast Guard should:

- identify, catalog, and establish communication with pertinent agencies, associations, groups, and individuals, both in government and in industry, at federal, regional, state, and local levels, in order to determine their respective current capabilities and future potential to function as part of a nationwide safety infrastructure network to assist in the development and conduct of the program;
- evaluate its maritime law enforcement program, including boardings and other compliance activities, to determine whether, to what extent, and how most effectively this program might be employed

in implementing a fishing industry vessel safety program to motivate as much as to demand compliance with safety regulations; and

- consider, as part of initial goal setting, each proposed safety improvement alternative in terms of required manpower, costs (including to whom), anticipated effectiveness, and implementation timing.

RECOMMENDATION
Upgrade Safety Data

The Coast Guard should upgrade safety data to provide the information needed to administer an integrated safety system. The Coast Guard should:

- assess fishing industry vessel safety data requirements, including data on fishing fleets and fishermen;
- consolidate, correlate, or otherwise provide compatibility between existing Coast Guard data bases and information systems, including the agency's main casualty (CASMAIN), search and rescue (SAR), and Summary Enforcement Event Report (SEER) data bases and the Marine Safety Information System (MSIS);
- expand and integrate data acquisition and utilization capabilities of these data bases in order to gather, standardize, evaluate, and disseminate fishing vessel safety data. The Coast Guard's Marine Accident Report Form, CG-2692, should be modified to include information on the fishery and activity within the fishery in which a commercial fishing industry vessel was engaged;
- coordinate with OSHA, NOAA, state offices maintaining vital statistics and casualty data, and the commercial fishing and marine insurance industries, within their functional areas of responsibilities, to further develop and integrate data on commercial fishing industry vessel casualties, fatalities, and injuries;
- upgrade existing federal and state vessel-registration programs to develop a comprehensive national data base encompassing *all* commercial fishing industry vessels for regulatory tracking purposes and to improve future analytic capabilities. The data should provide a basic record of vessel usage, details of the vessel's physical characteristics, and the nature of its employment;
- coordinate with NOAA and state agencies maintaining fishery license or permit data to develop a comprehensive national data base encompassing all fishermen to provide a basic record of the population at risk and for the purpose of improving analytical capabilities. The Coast Guard should establish a mandatory professional registration requirement if necessary to derive this information; and
- publish an annual report on fishing industry vessel safety, including information on vessel loss, fatality, and injury rates by region

and fishery. This annual report should include occupational safety data for the commercial fishing industry harvest sector comparable to those available for other industries. The report should provide the data necessary for evaluating the effectiveness of national fishing vessel safety efforts.

These recommendations correspond to alternatives 1 to 3, 7, and 15, which were introduced in previous chapters and are summarized in Table 8-1 with all other alternatives.

NEAR-TERM SAFETY IMPROVEMENT OPTIONS

There is a broad range of alternatives that can potentially improve safety. Each alternative needs to be evaluated in the context of an integrated safety program. The less-costly ones and those capable of early implementation deserve priority consideration. The committee's evaluation of implementation timing and applicability is also shown in Table 8-1. Some alternatives of moderate cost are already in partial use, providing a means for proceeding with safety improvements at a level industry could bear while the potential effects, benefits, and costs of more-stringent alternatives are evaluated. The near-term alternatives that should be employed as initial elements of an integrated safety improvement program follow.

Fishing Industry Vessels

Overall vessel fitness for service needs to be improved. The high incidence of breakdowns and casualties associated with material failure is indicative of widespread deficiencies throughout the national fishing fleet. Even many vessels with minor deficiencies that continue their pattern of minor breakdowns present, under certain conditions, the potential for major casualties affecting both lives and vessels. Reducing safety deficiencies for vessels not fully fit for service should reduce casualty rates. The existing data do not permit identification of specific vessels at risk. Therefore, improvement measures would initially have to be applied across the entire fleet to be effective and would include some form of compulsory inspection. The inspection could be as benign as a mandatory self-inspection backed by suitable compliance measures. If this failed to achieve desired objectives, a series of more-thorough and -costly technical alternatives would culminate in a formal vessel-inspection program. Some vessels are likely to be unfit to continue operations in any capacity and should be retired or constrained from further service. Removal from service would occur to some degree if load-line, classification, or full inspection requirements were imposed.

The character of commercial fishing is changing with the advent of increasing numbers of larger vessels, expansion of U.S. fish processing activity

TABLE 8-1 Safety Improvement Alternatives

		IMPLEMENTATION			
Alt. Number	**Category/Alternative Title**	**Near-term**	**Mid-term**	**Long-term**	**Not carried forward to a recommendation**
	SAFETY ADMINISTRATION				
1	**Update and expand safety data**	♦			
2	Require vessel registration	x			
3	Require professional registration	o			
	VESSEL-RELATED ALTERNATIVES				
4	**Establish minimum design, structural, stability, and material condition standards**	x			
5	**Expand equipment requirements**	♦			
6	Improve human engineering of vessels, deck layouts, and machinery			♦	
7	Continue compliance examinations	♦			
8	Require self-inspection	♦			
9	**Require marine surveys**		x		
10	Require load lines			x	
11	**Require vessel classification**			x	
12	Require **vessel inspection**		x		
13	**Remove unsafe**, inefficient, or excess **vessels from service**		x		
	PERSONNEL-RELATED ALTERNATIVES				
14	Establish risk communication/safety awareness programs	o			
15	Publish and distribute safety publications	o			
16	Require emergency preparedness measures	o			
17	Develop and promulgate standard operating procedures	o			
18	Develop competency standards	o			
19	Promote education and training	o			
20	Require education and training with certification		o		
21	Require **licensing**			o	
22	Establish vessel manning and watchkeeping criteria			o	
	SURVIVAL ALTERNATIVES				
23	Require manufacturers to provide installation, maintenance, and use instructions	o			
24	Develop and require carriage of fishing-industry-specific survival equipment	o			
25	Prohibit use of survival equipment that is not Coast Guard-approved				o
	EXTERNAL INFLUENCES ALTERNATIVES				
26	Establish flexible season openings	o			
27	Establish a voting position for a marine safety organization on each fishery management council		o		
28	Expand safety emphasis of fishery management plans			♦	
29	Require insurance coverage				♦
30	Expand fishing-industry-specific weather services	o			

Primary applicability: x - vessels; o - personnel; ♦ - vessels and personnel; **Boldface** - specifically addressed in CFIVSA afloat, and technologically more-advanced fishing gear. Some processing vessels already have significant complements of industrial workers beyond those traditionally described as fishermen. The importance of introducing industrial safety measures common in other industries will grow as more of these vessels join the fishing fleet.

RECOMMENDATION
Establish Vessel and Equipment Standards

The Coast Guard should establish minimum standards for vessel design, construction or conversion, arrangements, materials, and stability and should establish or expand carriage and maintenance requirements for navigation, communication, fire-fighting, and lifesaving equipment. These requirements should be correlated with vessel physical characteristics and usage and operating areas. The standards should be consistent with existing voluntary guidelines that demonstrably improve safety, and should be made mandatory for all new construction and conversions.

RECOMMENDATION
Utilize Regulatory Enforcement Activities

The Coast Guard should continue compliance examinations at an appropriate level to motivate adherence to safety regulations, modifying the scope and level of enforcement in consultation with the fishing industry as other alternatives are applied to safety problems.

RECOMMENDATION
Require Inspection

The Coast Guard should establish and administer regulations requiring a compulsory self-inspection program to improve vessel fitness for intended service. The program should contain:

- a methodology through which owners and operators of fishing industry vessels, not subject to more-stringent inspection measures by other regulations, would conduct a self-inspection of their vessels in advance of a fishing season or extended voyage utilizing a prescribed checklist or other inspection guide to determine that the vessel is fit for service in accordance with standards and equipment regulations;
- an audit process, such as dockside or underway boardings, other form of compliance examinations, or reporting regime, through which self-inspection can be validated and confirmed;
- provisions for accepting more-thorough examinations, such as a marine survey by a qualified third party, vessel classification, or maintenance in class in lieu of self-inspection;

- provisions for imposing more-stringent inspections or sanctions on a vessel-by-vessel basis by the auditing agency or its representative on a finding of excessive or unresolved discrepancies or other determination that a vessel is not being fully or properly maintained; and
- provisions for advancing to more-stringent inspection alternatives for some or all vessels if self-inspection proves unsatisfactory or ineffective in improving safety.

RECOMMENDATION
Remove Unfit Vessels from Service

The Coast Guard, in consultation with the National Marine Fisheries Service (NMFS), should research the merit of safety and economic programs for permanently removing vessels no longer fit for service from the U.S. fishing fleet.

RECOMMENDATION
Improve Safety in the Workplace

The Coast Guard, in concert with OSHA, should research ways in which occupational safety in the marine environment could be improved for activities of an industrial nature aboard fishing industry vessels.

These recommendations address alternatives 4 through 8 and 13. If vessel self-inspection, alternative 8, fails to achieve its objectives within a reasonable period developed in consultation with the fishing industry, consideration should then be given to independent surveys, load-line requirements, classification, or formal inspection, alternatives 9 through 12.

Human Factors

There are no data that permit determination of the professional experience or qualifications of individual fishermen, including vessel operators and watchkeepers, or their awareness of or attention to safety. However, the prevalence of human factors as direct or indirect causes of vessel and personnel casualties and personal injuries in the data, and rich anecdotal information, indicate that safety problems related to the human dimension are widespread. The casualties that occur are frequently associated with insufficient awareness of safety issues and basic expertise, which can be mitigated or compensated for by:

- increasing awareness of safety as a fundamental responsibility of owners, operators, and crewmen, not only in their self-interest, but as an element of good business;

- providing reasonable means for all fishermen and vessel operators to acquire the basic skills needed to successfully perform their respective roles; and
- ensuring that the basic qualifications needed for service as vessel operators, watchkeepers, and crewmen are attained.

Some form of training with validation of competency will be needed for all fishermen and vessel operators. Measures could be as benign as onboard instruction with log entry verifications and local-level training programs with certificates of completion. If these failed to achieve the desired objectives, more-thorough licensing regimes for fishermen and operators should culminate in formal license examinations. Measures to improve basic skills could be started by encouraging use of existing training resources, expanding this capability and accrediting curricula and training facilities, and then requiring participation in training programs. Program effectiveness would be enhanced by providing training opportunities at fishing ports and by correlating scheduling with fishing operations and seasons. Different training levels are needed for each role. Verification of training is needed and could be accomplished by issuing certificates from approved training facilities, for presentation on demand by the auditing organization. Improved accountability for safety is needed and could be accomplished by requiring vessel operator licenses issued upon completion of training and examination requirements.

RECOMMENDATION
Expand Safety Awareness

The Coast Guard, in conjunction with the fishing industry, Maritime Administration (MARAD), NOAA, and OSHA, should organize and lead an intensive, broad-based risk communication effort to improve safety awareness among members of the fishing industry. The program should be aimed at informing, educating, and motivating fishermen on matters of safety and its impact on their lives and livelihoods.

RECOMMENDATION
Improve Emergency Preparedness

The Coast Guard should immediately establish regulations requiring basic emergency preparations by all personnel aboard fishing industry vessels. The regulations should mandate onboard safety orientation, instructions, and emergency drills. The Coast Guard should, in consultation with NOAA and the fishing industry, develop user-friendly materials and methodologies to facilitate compliance.

RECOMMENDATION
Establish Basic Professional Qualification Standards

The Coast Guard, in conjunction with the fishing industry, should identify the minimum basic qualification levels needed for all persons engaged in commercial fishing and the standard operating procedures that should be employed. The Coast Guard should publish and encourage use of standard operating procedures (including manning and watchkeeping guidance), insofar as practical, in the fishing fleet.

RECOMMENDATION
Enhance the Education and Training Infrastructure

The Coast Guard, in conjunction with MARAD and NOAA, should enhance the existing education and training infrastructure, including development of accreditation standards and establishment of a sufficient national, regional, and local resource base, to ensure the means through which fishermen can obtain basic knowledge and practical skills as crewmen, watchkeepers, and operators. The Coast Guard, NOAA, and fishing industry leaders should encourage use of existing training opportunities to acquire basic knowledge and skills.

RECOMMENDATION
Require Professional Competency

The Coast Guard should establish and administer regulations requiring that each fisherman, vessel operator, or individual in charge acquires the fundamental skills associated with his or her role aboard fishing industry vessels, as follows:

- The Coast Guard should establish a certification program to provide a means for each fisherman to establish his or her basic qualifications for employment in the industry by meeting criteria tailored for the industry, such as time in service, attendance at educational or training courses, or demonstrations of competence.
- The Coast Guard should establish a licensing requirement applicable to each operator or individual in charge of a fishing industry vessel. Implementation of the license requirement should emphasize development of the practical skills needed to operate different categories of fishing industry vessels while also providing the means for holding operators accountable for safety. The operator license should be issued upon presentation of a certificate of competency, acceptable to the Coast Guard, attesting to satisfactory completion of the required courses pertaining to vessel operation and safety, except where the existing license for master or mate of an uninspected fishing industry vessel is required or held.
- The Coast Guard should establish an audit process such as

verification through boardings or a professional registration program, employing automated data bases for effective information management, to ensure that fisherman certification and operator licensing requirements are met.

- If performance objectives are not met through measures intended to facilitate skill development at the local level, the Coast Guard should establish provisions for advancing to more-stringent licensing measures for fishermen and vessel operators, such as requiring formal examinations and mandating manning and watchkeeping requirements.

These recommendations correspond to alternatives 3 and 14 through 22.

Survival

The effectiveness of available Coast Guard-approved survival equipment would be enhanced by providing improved instructional material specifically oriented toward its maintenance and use in the fishing environment. Coast Guard-approved, special-purpose deckwear with inherent or inflatable flotation and thermal protection suitable for use aboard small fishing vessels is not available. Equipment of this type is needed to provide fishermen a reasonable way to protect themselves against falls overboard and sudden, catastrophic loss of their vessels that precludes access to or use of other emergency equipment.

RECOMMENDATION
Improve Use and Maintenance Instructions for Survival Equipment

The Coast Guard should require that each item of Coast Guard-approved, special-purpose survival equipment be accompanied by adequate instructional material, including audiovisual aids, demonstrating correct use and maintenance to assist fishermen in improving the readiness of survival equipment and their ability to effectively employ this equipment in survival settings.

RECOMMENDATION
Improve Special-Purpose Survival Equipment

The Coast Guard should, in consultation with the commercial fishing industry, identify special-purpose equipment specifically designed for use aboard fishing vessels that is needed to increase the likelihood that fishermen will survive falls overboard or sudden loss of their vessel, develop standards for this equipment, and develop prototype equipment if necessary to bring this equipment to market.

> The Coast Guard should consider the merit of requiring the carriage of such equipment after a thorough field evaluation.

These recommendations correspond to alternatives 23 and 24. Insufficient data are available to make a preliminary determination on the merit of alternative 25, to prohibit nonapproved survival equipment.

Fisheries Management

Fishery management practices are not direct causes of casualties in the fishing industry. Nevertheless, they have contributed to the sequence of events leading to some casualties. Safety could be improved for short-opening fisheries by providing flexibility to accommodate poor environmental conditions. Attention to safety could be improved by establishing a neutral, voting member to represent safety issues on the regional fishery management councils. Economic conditions favorable to investment in safety equipment and training could be fostered by measures to match the national harvesting capability with the available fishery resources.

RECOMMENDATION
Increase Attention to Safety as an Element of Fisheries Management

> The Secretary of Transportation and the Under Secretary of Commerce for Oceans and Atmosphere should petition Congress to establish a Coast Guard flag officer as a voting member on each of the fishery management councils and to add safety considerations to national standards stated in the Magnuson Fisheries Conservation and Management Act for the express purpose of establishing safety as an equal consideration with other factors in fisheries management decision making.

This recommendation corresponds to alternatives 26 through 28. Management of harvesting capacity, an element of alternatives 13 and 28, requires a comprehensive economic analysis which was beyond the scope of this study and is not carried forward to a specific safety recommendation.

Insurance

Insurance does not cause casualties in the fishing industry—it pays for them. Insurance rates are a direct reflection of the industry's unsatisfactory safety record. Nevertheless, making insurance compulsory, alternative 29, is not practical under the present structure and economic conditions of the fishing and insurance industries, and it is not clear that the alternative could be implemented in such as way as to have the desired effect of improving safety.

Weather Services

Weather advisory services are not implicated as direct causes of casualties in the fishing industry. However, weather conditions are clearly the proximate cause of some casualties and contribute to many more. The availability of timely, accurate, and complete weather information for the fishing grounds and fishing ports, particularly those that are remote or prone to rapidly changing conditions, potentially would improve the opportunity for timely decision making by vessel operators.

RECOMMENDATION
Improve Weather Services

The National Weather Service should research fishing industry weather advisory needs to determine if additional coverage is needed for fishing grounds and ports and, if needed, take action necessary to provide such services.

This recommendation corresponds to alternative 30.

SUMMARY

The recommendations propose a single, integrated program for safety improvement under Department of Transportation leadership that would begin immediately by expanding existing measures and drawing on the existing safety infrastructure and resources. In this way, government, industry, and individual resources would not be unduly strained, nor further delay experienced, in establishing systematic attention to safety.

Treating safety as a total concept does not mean that all elements of the system have to be given the same priority or activated concurrently. It does mean, however, seeking an effective balance among all program elements to maximize the effective contribution of incremental costs and cumulative impacts of each.

Since it is not known how effective individual alternatives might prove to be in application, it makes sense to begin with basic measures to address each major problem area—safety performance monitoring, vessel-related problems, personnel-related problems, survival issues, and external influences. Basic alternatives in each of these areas could be refined and given appropriate emphasis as experience is gained during application. The alternatives identified in this report, consolidated in Table 8-1, are categorized by problem area and the estimated general opportunity for implementation. It is envisioned that near-term items could form the foundation for a comprehensive program encompassing most or all of the alternatives shown.

Data that are now fragmented, incomplete, or nonexistent could begin to be organized and improved through the proposed registration and improved casualty-reporting systems. This would lead to a data resource of meaningful potential for administration of a fully integrated safety program. Owners and operators should be provided the opportunity through improved safety performance to avoid the more extreme, costly alternatives, such as a full vessel inspection requirement. If, however, the less rigorous measures do not achieve the desired objectives in a reasonable period, the safety program should be stepped up over the long term until the desired level of performance is achieved.

Setting goals and objectives and measuring progress will be a challenge. At each stage, the following questions must be asked: How much safety is enough? What costs is the industry able to bear? And how many resources is government willing to devote to fishing vessel safety? Ultimately, the level of federal and industry resources that can be committed to improving safety will be a principal determinant of the configuration of the resulting programs.

References

Adee, B. H. 1985. A review of some recent stability casualties involving Pacific Northwest fishing vessels. Pp. 217-237 in International Conference on Design Construction and Operation of Commercial Fishing Vessels—Proceedings, J. C. Sainsbury and T. M. Leahy, eds. Florida Sea Grant College. May 1985.

Adee, B. H. 1987. Vessel stability: knowledge = safety. Alaska Marine Resource Quarterly 2 (First Quarter):2-4.

Alaska Department of Labor. 1988. Occupational Injury and Illness Information, Alaska 1986. Juneau. May.

Alaska Fisheries Safety Advisory Council. 1977. Alaska Fisheries Safety Advisory Council Interim Fishing Vessel Safety Standards. Marine Advisory Bulletin No. 5, July 1977. University of Alaska, Fairbanks.

Alaska Fisherman's Journal. 1987. Halibut opening marred by losses. 10 (June):12.

Alaska Fisherman's Journal. 1988a. Don't trust your life to luck. 11 (May):11.

Alaska Fisherman's Journal. 1988b. The sinking of the Snow Mist. 11 (May):8-11.

Alaska Fisherman's Journal. 1989a. Fishermen drift four days after vessel sinks. 12 (November):48-49.

Alaska Fisherman's Journal. 1989b. Two dead, one missing. 12 (November):4.

Allen, D., and D. W. Nixon. 1986. Pp. 58-86 in Study on the Use of Fishery Management Regulations and Techniques to Improve the Safety of Commercial Fishing Operations, T. J. Lassen and K. Van Olst, eds. National Council of Fishing Vessel Safety and Insurance, Washington, D.C.

Alverson, D. L. 1985. U.S. fisheries policy evolution. Pp. 3-17 in Proceedings of the Conference on Fisheries Management: Issues and Options, T. Frady, ed. Alaska Sea Grant Report 85-2, Fairbanks.

Amagai, K., N. Kimura, and T. Oosaka. 1989. A proposal to prevent occupational accidents in scallop beam trawler. Paper presented at International Symposium: Safety and Working Conditions Aboard Fishing Vessels, l'Université du Québec à Rimouski, Quebec, Canada, August 22-24, 1989.

American Bureau of Shipping (ABS). 1989. Guide for Building and Classing Fishing Vessels. Paramus, N.J.: ABS.

American Hull Insurance Syndicate (AHIS). 1964-1986. Reports of Annual Meetings of Subscribers to the American Hull Insurance Syndicate for Years 1964 Through 1986. New York, N.Y.: AHIS.

American Institute of Marine Underwriters. 1985. Statement of the American Institute of Marine Underwriters. Pp. 81-94 in Fishing Vessel Insurance—Part 1. Serial No. 99-28, U.S. Congress, House of Representatives. Washington, D.C.: Government Printing Office.

American Institute of Marine Underwriters. 1988. 1988 Annual Report. New York. P. 5.

American Society for Testing and Materials (ASTM). 1988. Standard Practice for Human Engineering Design for Marine Systems, Equipment and Facilities. Philadelphia. P. 155.

Appave, M. D. 1989. The role of the ILO in the improvement of safety and working conditions aboard fishing vessels. Paper presented at International Symposium: Safety and Working Conditions Aboard Fishing Vessels, l'Université du Québec à Rimouski, Quebec, Canada, August 22-24, 1989.

Ball, J. 1978. Vessel icing forecasts: part of winter fishing safety. (Originally printed in Alaska Seas and Coasts, December 1978). In Readings from . . . Alaska Seas and Coasts. Alaska Sea Grant Report 79-5, August 1979, Fairbanks.

Bard, R. 1990. Avoiding collisions while drifting at sea. National Fisherman 71 (May):10-12.

Bárdarson, H. R. 1984. Safety of fishing vessels and their crew. IMO News 4:2-3.

Bernton, H. 1990. Trawling free-for-all threatens Alaska catch. Washington Post, May 26.

Best's. 1988. Best's Aggregates and Averages, 1988. Oldwick, N.J.: A. M. Best.

Blewett, E., W. Furlong, and P. Toews. 1985. Canada's experience in measuring the deterrent effect of fisheries law enforcement. Paper presented at International Workshop on Fisheries and Law Enforcement, University of Rhode Island, October 21-23, 1985.

Boehmer, K. 1989. Postscript: Excerpt from a letter by Kris Boehmer, reflecting on the loss of the Janileen II. (Originally printed in Commercial Fisheries News, October 1989.) In E. Yoder, For those in peril on the sea: progress in commercial fishing safety. Nor'Easter 2 (Spring 1990):13.

Booth, R. 1990. Winter storm claims the lives of five lobstermen. National Fisherman 71 (February):16.

Bourke, R. 1990. Regional assessment for Hawaii and distant-waters fisheries region. Unpublished contract report to the Marine Board, National Research Council, Washington, D.C.

Browning, R. J. 1980. Fisheries of the North Pacific: History, Species, Gear, and Processes. Rev. ed. Anchorage, Alaska: Alaska Northwest Publishing. 423 pp.

Buls, B. 1990. A new breed—factory longliners for the North Pacific. National Fisherman 70 (January):44-45.

Campbell, T. 1990. Vessel anti-reflagging law attacked as weak. National Fisherman 71 (May):22-23.

Canada, Government of, Minister of Labour. 1988. Report of the Committee on Occupational Safety and Health in the Fishing Industry. Ottawa: Labour Canada.

Canadian Coast Guard. 1987. A Coast Guard Study into Fishing Vessel Safety. Prepared for Canadian Coast Guard by Coast Guard Working Group on Fishing Vessel Safety. October. Ottawa: Transport Canada.

Canadian Coast Guard. 1988. A Training Programme in Marine Emergency Duties: Canadian Coast Guard Ship Safety 1987 (Revised June 1988). Ottawa: Transport Canada.

Canadian General Standards Board. 1989. National Standard of Canada: Marine Anti-Exposure Work Suit Systems. Ottawa.

Carbajosa, J. M. 1989. Report on work accidents in the sea fishing sector. Paper presented at International Symposium: Safety and Working Conditions Aboard Fishing Vessels, l'Université du Québec à Rimouski, Quebec, Canada, August 22-24, 1989.

Carter, J. M. 1989. Federal/provincial initiatives on occupational safety and health in the fishing industry. Paper presented at International Symposium: Safety and Working Conditions Aboard Fishing Vessels, l'Université du Québec à Rimouski, Quebec, Canada, August 22-24, 1989.

Castle, R. D. 1988. F/V Wayward Wind: sinking with the loss of four lives. In Regional assessment for Alaska region, N. Munro, 1990, unpublished contract report to the Marine Board, National Research Council, Washington, D.C.

Chandler, A. D. 1988. The National Marine Fisheries Service. Audubon Wildlife Report 1988/1989, W. J. Chandler, L. Labate, and C. Wille, eds. New York: Academic Press.

Clendinen, D. 1984. As fishing industry changes, risk and loss seem to grow. New York Times, December 10. P. 1.

Colucciello, L. A. 1988. Unsafe shipping and polluted oceans: marine casualties and marine casualty investigations. Paper presented at Fifth International Conference on Maritime Education and Training, Sydney, Nova Scotia, Canada, September 19-22, 1988.

Commercial Fishing Industry Vessel Advisory Committee. 1989. Minutes of Commercial Fishing Industry Vessel Advisory Committee, Third Meeting, October 22-24, 1989, Seattle, Washington.

Cooper, C. 1989. How did the shrimping industry get TEDS? National Fisherman 70 (November):18-19.

Coopers and Lybrand. 1990. Economic Impacts of the North Pacific Factory Trawler Fleet. Seattle, Washington. January.

Coughenower, D. D. 1986. Homer, Alaska, Charter Fishing Industry Study. Alaska Sea Grant College Program, Marine Advisory Bulletin No. 22. Fairbanks: University of Alaska.

Coughenower, D. D. 1987. Commercial Fishing Industry Study, Homer, Alaska. Alaska Sea Grant College Program, Marine Advisory Bulletin No. 33. Fairbanks: University of Alaska.

Couper, A. 1985. Social consequences of maritime technological change. Presented as Donald L. McKernan Lectures in Marine Affairs, November 26, 1985. Washington Sea Grant Program and Institute for Marine Studies. Seattle: University of Washington.

Dahle, E. A., and J. C. D. Weerasekera. 1989. Safe design and construction of small fishing vessels. Paper presented at International Symposium: Safety and Working Conditions Aboard Fishing Vessels, l'Université du Québec à Rimouski, Quebec, Canada, August 22-24, 1989.

Danforth, W. C., and C. A. Theodore. 1957. Hull Insurance and Protection and Indemnity Insurance of Commercial Fishing Vessels. U.S. Fish and Wildlife Service, Special Scientific Report—Fisheries Nos. 241 and 241 Supplement. Washington, D.C.: Department of the Interior.

Day, J. 1990. Volunteer rescue unit saves lives on Maine coast. National Fisherman 70 (January):23-25.

DeAlteris, J., R. Wing, and K. Castro. 1989. The fishing vessel safety training program at the University of Rhode Island, U.S.A. Paper presented at International Symposium: Safety and Working Conditions Aboard Fishing Vessels, l'Université du Québec à Rimouski, Quebec, Canada, August 22-24, 1989.

deCarteret, J. E., N. W. Lemley, and D. F. Sheehan. 1980. Life safety approach to fishing vessel design and operation. Paper presented at Spring Meeting/STAR Symposium, Coronado, California, June 4-5, 1980. New York: The Society of Naval Architects and Marine Engineers. Pp. 177-186.

Degener, R. 1989a. Hope fades for three lost at sea. Press of Atlantic City, December 17. Pp. A1, A10.

Degener, R. 1989b. Hughes to probe Coast Guard's efforts to aid clam boat. Press of Atlantic City, December 16. P. B1.

Degener, R., and G. Strawley. 1989. Fishermen, Coast Guard search seas off Cape May. Press of Atlantic City, December 15. Pp. A1, A4.

Dewees, C. M., and G. R. Hawkes. 1988. Technical innovation in the Pacific Coast trawl fishery: the effects of fishermen's characteristics and perceptions on adoption behavior. Human Organization 47 (3):224-234.

Dobravec, M. L. 1987. Lessons from casualties: working over the side may be hazardous to your health. Proceedings of the Marine Safety Council (June):145-147.

Dyer-Smith, M. B. A., and A. F. M. De Bievre. 1988. The human element in marine accidents. Paper presented at Fifth International Conference on Maritime Education and Training, Sydney, Nova Scotia, Canada, September 19-22, 1988.

Dynamics Research Corporation. 1989. The Role of Human Factors in Marine Casualties. Final contract report prepared for U.S. Coast Guard Headquarters. Arlington, Va.: Dynamics Research Corporation.

Eberhardt, P. T. 1989. An Investigation of the Roll Period Test for Small Fishing Vessels. Master's thesis. University of Washington.

Ecker, W. J. 1978. A safety analysis of fishing vessel casualties. Prepared for the Sixty-Sixth National Safety Congress and Exposition, Chicago, Ill., October 2-5, 1978.

Embler, J. V. Undated. EPIRBs 101. Unpublished U.S. Coast Guard staff paper. Search and Rescue Division, U.S. Coast Guard Headquarters, Washington, D.C.

Esbensen, P., R. E. Johnson, and P. Kayten. 1985. The importance of crew training and standard operating procedures in commercial vessel accident prevention. The Society of Naval Architects and Marine Engineers, Spring Meeting STAR Symposium, Norfolk, Va., May 21-24, 1985.

Expert. 1989. OSHA and fishing industry vessels. Maritime and Environmental Consultants, Ridgefield, Wash. (Fall):3-5.

Expert. 1990. Choosing a good surveyor witness. Maritime and Environmental Consultants, Ridgefield, Wash. (Spring):4-5.

Federal Register. 1990. Commercial Fishing Industry Vessel Regulations. 55 (76): April 19, 1990. Pp. 14924-14960.

Finley, C. 1982a. Man overboard! National Fisherman Yearbook. Pp. 40-42.

Finley, C. 1982b. When to call for help. National Fisherman Yearbook. Pp. 24-26.

Fishermen's News. 1988. Wild halibut opening brings CG assistance. (July):1.

Fitzpatrick, J. 1989. Fishing technology. Pp. 90-116 in Ocean Yearbook 8, E. M. Borgese, N. Ginsburg, and J. R. Morgan, eds. Chicago: The University of Chicago Press.

Frady, T., ed. 1985. Proceedings of the Conference on Fisheries Management: Issues and Options. Fairbanks: Alaska Sea Grant Report 85-2, April 1985. 429 pp.
Freeman, K. 1990. Getting a handle on the right glove. National Fisherman 71 (August): 52-53.
Gale, R. P. 1990. Fisherman types and fisherman safety. Attachment 1, Regional safety data: West Coast region, of the West Coast commercial fishing vessel safety assessment, prepared by R. Jacobson, G. Goblirsch, and F. Van Noy. Unpublished report to the Marine Board, National Research Council, Washington, D.C.
Gleason, R. 1982. How hazardous is harvesting the sea? A report on fishing fatalities. Proceedings of the Marine Safety Council (June):173-176.
Glickman, A. 1984. Insurers raise rates as claims rise for fishing vessels in Gloucester, Mass. Wall Street Journal, August 7. P. 8.
Gordon, W. G. 1989. U.S. commercial fisheries: a brief review. Unpublished contract report to the Marine Board, National Research Council, Washington, D.C.
Goudey, C. 1986a. Winch accident: it could happen to you. In Fishing Vessel Topics: Reprints from Commercial Fisheries News, C. A. Goudey. August 1988. Cambridge, Mass.: MIT Sea Grant Program.
Goudey, C. 1986b. Safety first for winch design. In Fishing Vessel Topics: Reprints from Commercial Fisheries News, C. A. Goudey. August 1988. Cambridge, Mass.: MIT Sea Grant Program.
Grasselli, N. E., and M. A. O'Hara. 1983. Coast Guard enforcement of fishery laws and treaties. Unpublished U.S. Coast Guard paper. September.
Gray, R. J. 1986. An examination of the occupational safety and health situation in the fishing industry in B.C. Unpublished report, West Vancouver, B.C.
Gray, R. J. 1987a. An examination of the occupational safety and health situation in the inland fishing industry of Canada. Unpublished report, West Vancouver, B.C.
Gray, R. J. 1987b. An examination of the occupational safety and health situation in the Atlantic fishery of Canada. Unpublished report, West Vancouver, B.C.
Gray, R. J. 1987c. An examination of the occupational safety and health situation in the fisheries of Canada. Unpublished report, West Vancouver, B.C.
Griffen, N. 1989a. East Coast factory trawler fleet is alive and well. National Fisherman 70 (November):14-16.
Griffen, N. 1989b. Scallopers start freezing at sea. National Fisherman 70 (November):16.
Griffen, N. 1990. Why the Northeast rejected automated longlining. National Fisherman 70 (April):44-45, 72.
Grissim, J. 1990. Probing the mysteries of giant waves. National Fisherman 71 (July):21-23.
Gross, E. 1958. Work and Society. New York: Thomas Crowell.
Hamilton, R. A. 1989. The rewards outweigh the risks for many who find a home at sea. New London Day, November 29.
Hasselback, P., and C. I. Neutel. 1990. Risk for commercial fishing deaths in Canadian Atlantic provinces. British Journal of Industrial Medicine 47:498-501.
Hatfield, P. S. 1989. Fishing vessel safety—stability considerations. Paper presented at International Symposium: Safety and Working Conditions Aboard Fishing Vessels, l'Université du Québec à Rimouski, Quebec, Canada, August 22-24, 1989.
Hockey, G. R. J. 1986. Changes in operator efficiency as a function of environmental stress, fatigue, and circadian rhythms. Pp. 44-1 to 44-49. In Boff, K. R., L. Kaufman, and J. P. Thomas, eds. Handbook of Perception and Human Performance, Vol. II. New York: John Wiley & Sons.

Hoefnagel, W. A. M., and K. Bouwman. 1989. Safety aboard Dutch fishing vessels. Paper presented at International Symposium: Safety and Working Conditions Aboard Fishing Vessels, l'Université du Québec à Rimouski, Quebec, Canada, August 22-24, 1989.

Hollin, D. 1982. Safety at Sea: A Guide for Fishing Vessel Owners and Operators. Prepared for Sea Grant College Program, Texas A&M University. September.

Hollin, D. 1990. Gulf of Mexico regional assessment. Unpublished contract report to the Marine Board, National Research Council, Washington, D.C.

Hollin, D., and E. D. Middleton, Jr., eds. 1989. Gulf Coast Fishing Vessel Safety Manual. Washington, D.C.: National Council of Fishing Vessel Safety and Insurance.

Hollister, B., and A. Carr, eds. 1990. Marine Living Resources Engineering and Technology. Washington, D.C.: Marine Technology Society.

Hopper, A. G., and A. J. Dean. 1989. Safety engineering and equipment for fishing vessels under 20m: results of trials on two vessels—Madalia and Silverline. Paper presented at International Symposium: Safety and Working Conditions Aboard Fishing Vessels, l'Université du Québec à Rimouski, Quebec, Canada, August 22-24, 1989.

IMO News. 1989. SFV convention review makes progress. 2:8.

IMO News. 1990. Priority to be given to fishing safety. 1:5.

International Maritime Organization (IMO). 1966. International Convention on Load Lines. London: IMO.

International Maritime Organization (IMO). 1975a. Code of Safety for Fishermen and Fishing Vessels: Part A—Safety and Health Practices for Skippers and Crews. London: IMO.

International Maritime Organization (IMO). 1975b. Code of Safety for Fishermen and Fishing Vessels: Part B—Safety and Health Requirements for the Construction and Equipment of Fishing Vessels. London: IMO.

International Maritime Organization (IMO). 1977. International Conference on Safety of Fishing Vessels 1977: Final Act of the Conference, with attachments, including the Torremolinos International Convention for the Safety of Fishing Vessels, 1977. London: IMO.

International Maritime Organization (IMO). 1980. Voluntary Guidelines for the Design, Construction and Equipment of Small Fishing Vessels. London: IMO.

International Maritime Organization (IMO). 1988. Document for Guidance on Fishermen's Training and Certification: An International Maritime Training Guide. London: IMO.

Jacobson, R., G. Goblirsch, and F. Van Noy. 1990. West Coast commercial fishing vessel safety assessment. Unpublished contract report to the Marine Board, National Research Council, Washington, D.C.

Jelvik, M. L. 1986. Washington State's experience with limited entry. Pp. 313-316 in Fishery Access Control Programs Worldwide: Proceedings of the Workshop on Management Options for the North Pacific Longline Fisheries, N. Mollet, ed., Alaska Sea Grant Report No. 86-4, University of Alaska, Fairbanks. December.

Johnson, T. 1982. Modern safety gear boosts your chances of surviving. National Fisherman Yearbook. Pp. 46-51.

Jones, R. P., Project Director. 1987. Loss Control and Risk Finance for the Gulf and South Atlantic Commercial Fishing Fleet. Prepared by Southeastern Fisheries Association, Inc., Tallahassee, Fla. June.

Jones, R. P. 1990. Regional assessment for Caribbean region. Unpublished contract report to the Marine Board, National Research Council, Washington, D.C.

Jones, R. P., and J. Maiolo. 1990. South Atlantic regional assessment. Unpublished contract report to the Marine Board, National Research Council, Washington, D.C.

Keiffer, E., ed. 1984. Fishing Vessel Safety Conference, 1983: Report on a Conference Held in Washington, D.C., November 9-10, 1983. The University of Rhode Island Marine Memorandum 76, prepared under NOAA Office of Sea Grant, Department of Commerce, Grant #NA81AA-D-00073 (Spring).

Kime, J. W. 1986. Statement of RADM. J. William Kime, Chief, Office of Merchant Marine Safety, U.S. Coast Guard. Hearings Before the Subcommittee on Merchant Marine of the Committee on Merchant Marine and Fisheries, House of Representatives, Ninety-Ninth Congress, on Limitation of Shipowners' Liability and Fishing Vessel Insurance and Safety Problems. April 17, 1986. Serial No. 99-36. Washington, D.C.: Government Printing Office.

Knapp, G., and N. Ronan. 1990. Fatality Rates in the Alaska Commercial Fishing Industry. Report No. 90-03. Fairbanks, Alaska: Alaska Sea Grant College Program.

Knox, S. A. 1990. Surveying fiberglass yachts. Marine Technology 27 (May):180-189.

Koch, C. L. 1985. Legal tools and restrictions affecting fisheries management. Pp. 149-184 in Proceedings of the Conference on Fisheries Management: Issues and Options, T. Frady, ed., Alaska Sea Grant Report 85-2, Fairbanks.

Lassen, T. 1985. Statement of Thor Lassen, Executive Secretary, National Council of Fishing Vessel Safety & Insurance. Pp. 5-11 in Fishing Vessel Safety and Insurance, S. Hrg. 99-268, U.S. Congress, Senate. Washington, D.C.: Government Printing Office.

Lassen, T. J., and K. Van Olst, eds. 1986. Study on the Use of Fishery Management Regulations and Techniques to Improve the Safety of Commercial Fishing Operations. Washington, D.C.: National Council of Fishing Vessel Safety and Insurance. 92 pp.

Lazarus, P. 1990a. Conn. lobster dealer convicted in sinking. National Fisherman 71 (August):10, 84.

Lazarus, P. 1990b. Buyers of new 406-MHz EPIRBs may experience delivery delays. National Fisherman 71 (August):33.

Lemon, D. 1990a. Status report on emergency beacons. Unpublished U.S. Coast Guard staff paper, Search and Rescue Division, U.S. Coast Guard Headquarters, Washington, D.C.

Lemon, D. 1990b. Why 406-MHz EPIRBs are a good catch. National Council of Fishing Vessel Safety and Insurance (April):2-3.

Lesh, T. 1982. Many fish boats lack adequate pumps, and bilge alarm systems are all too rare. National Fisherman Yearbook 1982. Pp. 32-35.

Letz, L. 1986. Scuttling doesn't pay. Proceedings of the Marine Safety Council (November/December). Pp. 256-257.

Levine, E. B., and B. J. McCay. 1987. Technology adoption among Cape May fishermen. Human Organization 46 (3):243-253.

Lucas, H. 1985. Operations. Pp. 135-154 in Official Transcript, Proceedings Before National Academy of Sciences, Marine Board, Symposium on Fishing Vessel Safety, Washington, D.C., December 11, 1985.

Lyon, G. H., and C. A. Theodore, eds. 1976. Summary Report of the Ad Hoc Group on Commercial Fishing Vessel Insurance, January 1973-May 1975. Prepared for National Oceanic and Atmospheric Administration. Washington, D.C.: Department of Commerce.

Macleod, I. K., and C. J. MacFarlane. 1989. Safety and its Implications for Design in the Scottish inshore fishing fleet. Paper presented at International Symposium: Safety and Working Conditions Aboard Fishing Vessels, l'Université du Québec à Rimouski, Quebec, Canada, August 22-24, 1989.

Maiolo, J. 1990. Memorandum prepared for Committee on Fishing Vessel Safety. Subject: Revised statement on the culture and social organization of commercial fishing. January 22, 1990.

Maiolo, J. R., and M. K. Orbach, eds. 1982. Modernization and Marine Fisheries Policy. Ann Arbor, Mich: Ann Arbor Science. 330 pp.

Marine Advisory Service. 1986. Impact of Fishery Management Methods on the Safety of Commercial Fishing Vessels and Their Crews—Selected Gulf of Mexico Fisheries. Report prepared by Marine Advisory Service, Sea Grant College Program, Texas A&M University, for the National Council of Fishing Vessel Safety and Insurance. August.

Matsen, B. 1989. Alaska's wild bottomfish ride nears its peak. National Fisherman 70 (November):89-92, 140.

Matsen, B. 1990. 162' Aleutian Enterprise capsizes, nine crewmen lost. National Fisherman 71 (June):11-13.

McCay, B. J., W. G. Gordon, E. B. Levine, J. B. Gatewood, B. Thompson, and C. Creed. 1989. From the Waterfront: Interviews with New Jersey Commercial Fishermen About Marine Safety and Training. Prepared for New Jersey Marine Sciences Consortium, National Marine Fisheries Service, Saltonstall-Kennedy Project for Understanding and Teaching Marine Safety.

McDowell, E., J. Calvin, and N. Gilbertson. 1989. Alaska Seafood Industry Study: A Technical Report. Prepared for the Alaska Seafood Industry Study Commission, Anchorage, Alaska.

McDowell Group. 1989. Alaska Seafood Industry Study: A Summary. Prepared for the Alaska Seafood Industry Study Commission, Anchorage, Alaska.

McDowell Group. 1990. The Effectiveness of Marine Safety Training. Prepared for the Alaska Marine Safety Education Association, Sitka, Alaska.

McGuffey, D. B., and J. C. Sainsbury. 1985. The effect of loading on fishing vessel stability: a case study. In J. C. Sainsbury, and T. M. Leahy, eds. International Conference on Design Construction and Operation of Commercial Fishing Vessels—Proceedings. Florida Sea Grant Program, May 1985. Pp. 289-310.

Melteff, B. R., Symposium Coordinator. 1988. Summary Proceedings of the National Workshop on Fishing Vessel Insurance and Safety. University of Alaska, Fairbanks, Alaska Sea Grant Report No. 88-2. May.

Mercy, D. 1990. Frostbite and Other Cold Injuries. Education Publication No. 7. Alaska Sea Grant College Program. Fairbanks, Alaska: University of Alaska. 19 pp.

Miller, C. 1985. Survival at sea: tips from some who've been there. Alaska Fisherman's Journal (November):10-11, 23.

Miller, G., and G. E. Miller. 1990. Human engineering applications to marine engineering systems. Paper presented at the 1990 Chesapeake Marine Engineering Symposium, Arlington, Va. March 14, 1990. Pp. 2-1 to 2-8.

Miller, M. L., and J. Van Maanen. 1982. Getting into fishing: observations on the social identities of New England fishermen. Urban Life 11(1).

Mohl, B. A. 1984. Insurance rates soar on Gloucester fishing boats. Globe Newspaper Company 226 (July 17):1.

Mollet, N. 1986. Fishery Access: Control Programs Worldwide. Proceedings of the Workshop on Management Options for the North Pacific Longline Fisheries. Fairbanks, Alaska: Alaska Sea Grant College Program. 366 pp.

Moran, R. 1989. Software program designed to monitor fishing vessel safety. Press release for National Council of Fishing Vessel Safety and Insurance. March.

Munro, N. 1990. Regional assessment for Alaska region. Unpublished contract report to the Marine Board, National Research Council, Washington, D.C.

Murray, J. J. 1962. Safety Manual for Fishermen, Captains, and Owners of New England Fishing Vessels. Washington, D.C.: Department of the Interior.

Myers, E. 1982. Don't kick those boots away. National Fisherman Yearbook. P. 44.

Nalder, E. 1990. Seas of neglect: 9 die after fishing safety law was slighted by firm and unenforced by the Coast Guard. Seattle Times-Seattle Post Intelligencer, November 18, 1990. Pp. A1, A16-A17.

National Council of Fishing Vessel Safety and Insurance. 1989. Safety Matters. December.

National Fisherman Yearbook. 1982. Seamanship figures prominently in how one deals with the weather. Pp. 71-72, 74, 191.

National Oceanic and Atmospheric Administration (NOAA). 1986. NOAA Fishery Management Study. Washington, D.C.: Department of Commerce.

National Oceanic and Atmospheric Administration (NOAA). 1989. Fisheries of the United States, 1988. Washington, D.C.: Department of Commerce.

National Oceanic and Atmospheric Administration (NOAA). 1990a. Fisheries of the United States, 1989. Washington, D.C.: Department of Commerce.

National Oceanic and Atmospheric Administration. 1990b. Fishing Trends and Conditions in the Southeast Region, 1989. Miami, Fla: National Marine Fisheries Service, Southeast Fisheries Center, Department of Commerce.

National Research Council (NRC). 1976. Human Error in Merchant Marine Safety. Washington, D.C.: National Academy of Sciences.

National Research Council. 1981. Research Needs to Reduce Maritime Collisions, Rammings, and Groundings. Washington, D.C.: National Academy Press.

National Research Council. 1985. Injury in America. Washington, D.C.: National Academy Press. 164 pp.

National Research Council. 1989a. Opportunities to Improve Marine Forecasting. Washington, D.C.: National Academy Press.

National Research Council. 1989b. Improving Risk Communication. Washington, D.C.: National Academy Press.

National Research Council. 1990. Decline of the Sea Turtles: Causes and Prevention. Washington, D.C.: National Academy Press.

National Transportation Safety Board. 1987. Safety Study: Uninspected Commercial Fishing Vessels, NTSB/SS-87/02. Springfield, Va.: National Technical Information Service.

National Transportation Safety Board. 1988a. Marine Accident Report: Capsizing and Sinking of the U.S. Fishing Vessel Lard Atlantic Ocean Near Nantucket Island, Massachusetts, October 9, 1987, NTSB/MAR-88/05. Springfield, Va.: National Technical Information Service.

National Transportation Safety Board. 1988b. Marine Accident Report: Capsizing and Sinking of the U.S. Fishing Vessel Uyak II in the Gulf of Alaska Near Kodiak Island, Alaska, November 15, 1987, NTSB/MAR-88/08. Springfield, Va.: National Technical Information Service.

National Transportation Safety Board. 1989a. Marine Accident Report: Sinking of the U.S. Fishing Vessel Wayward Wind in the Gulf of Alaska, Kodiak Island, Alaska, January 18, 1988, NTSB/MAR-89/01. Springfield, Va.: National Technical Information Service.

National Transportation Safety Board. 1989b. Safety Recommendation M-89-1 through -5. Washington, D.C.

National Transportation Safety Board. 1989c. Safety Recommendation M-89-6 through -8. Washington, D.C.

National Transportation Safety Board. 1989d. Safety Recommendation M-89-9 and -10. Washington, D.C.

National Transportation Safety Board. 1989e. Marine Accident Reports: Brief Format Issue Number 7—Reports Issued April 11, 1989, NTSB/MAB-89/01. Springfield, Va.: National Technical Information Service.

Natural Resource Consultants. 1986. Commercial Fishing and the State of Washington: A Contemporary Economic Overview of Local and Distant Water Fisheries. Prepared for a consortium of fishing industry organizations. Seattle, Wash.

Natural Resource Consultants. 1988. Commercial Fishing and the State of Washington: A Brief Overview of Recent and Future Growth in the Washington Seafood Industry. Prepared for a consortium of fishing industry organizations. Seattle, Wash.

Natural Resource Consultants. 1989. Recent Growth and Associated Impacts of the Washington and King County North Pacific Commercial Fishing Fleet. Prepared for The Port of Seattle. February.

Naughton, J. 1990. Eulogy for a death on the Western Sea: a bereft family's campaign to upgrade fishing safety. Washington Post, June 7. Pp. C1, C16.

Nies, T. A. 1986. Coast Guard fisheries law enforcement. Paper presented to Conference on East Coast Fisheries Law and Policy, Marine Law Institute, Portland, Maine, June 17-20, 1986.

Nies, T. A., and M. A. Carney. 1985. Problems of administering enforcement. Paper presented to Workshop on Fisheries Law Enforcement, University of Rhode Island, October 21-23, 1985.

Nixon, D. W. 1985. Statement of Dennis W. Nixon, Coordinator, Marine Affairs Program, University of Rhode Island, Kingston. Pp. 46-60 in Fishing Vessel Safety and Insurance, S. Hrg. 99-268, U.S. Congress, Senate. Washington, D.C.: Government Printing Office.

Nixon, D. W. 1986. Recent developments in U.S. commercial fishing vessel safety, insurance, and law. Journal of Maritime Law and Commerce 17(3):359-387.

Nixon, D. W. 1990. Regional assessment for the Northeast region. Unpublished contract report to the Marine Board, National Research Council, Washington, D.C.

Nixon, D. W., and F. M. Fairfield. 1986. Fishermen's personal injuries: characterization, compensation, and solutions. Paper for Graduate Program in Marine Affairs, University of Rhode Island.

Nixon, D., R. Moran, and C. Philbrick, eds. 1987. Self-Insurance Programs for the Commercial Fishing Industry. Workshop Summary Report for Alaska Sea Grant College Program, University of Alaska. Marine Advisory Bulletin No. 31. July.

NYNEX Information Resources Company. 1989. Marine Information. NYNEX Commercial Marine Directory: North Pacific Edition. Wakefield, Mass.

Offshore Research Focus. 1989. Immersion suits guidance published. August, p. 6.

Pacific Fisheries Consultants. 1987. National Workshop on Fishing Vessel Insurance and Safety, Washington, D.C., February 4-6, 1987: Final Report Prepared for Sea Grant Program, University of Hawaii, April 9, 1987.

Page, D. 1989. Family fishing—The joy of owning two boats. Pacific Fishing (May):52-56.
Parker, D. 1990. Kodiak fishermen defend the derby. Alaska Fisherman's Journal 13 (March):19-20.
Pawlowski, B. 1987. Emergency rescue: Sarsat beacons. Alaska Marine Resource Quarterly 2 (First Quarter):11.
Pennington, H. 1987. Alaska Marine Safety Education Association. Alaska Marine Resource Quarterly 2 (First Quarter):16
Piatt, C. 1990. New England trawler converted to squid catcher/processor. National Fisherman 71 (November):62-63.
Piche, G. G. 1985. Fishing safety analysis—U.S. Coast Guard. Pp. 92-105 in Official Transcript Proceedings Before National Academy of Sciences, Marine Board, Symposium on Fishing Vessel Safety, Washington, D.C., December 11, 1985.
Piche, G. G., W. J. Morani, Jr., and H. A. Chatterton, Jr. 1987. U.S. Coast Guard's fishing vessel safety initiative. Paper presented to the Chesapeake Section of the Society of Naval Architects and Marine Engineers, Crystal City, Va., March 25, 1987.
Pizzo, J. T., and S. Jaeger. 1974. An Overview of Commercial Fishing Vessel Safety in the Northwest and Alternatives for Loss Prevention. Seattle, Wash.: National Marine Fisheries Service, NOAA. 93 pp.
Plaza, F. 1989. IMO's work relating to fishing vessels. Paper presented at International Symposium: Safety and Working Conditions Aboard Fishing Vessels, l'Université du Québec à Rimouski, Quebec, Canada, August 22-24, 1989.
Poggie, J. J., and R. B. Pollnac. 1988. Danger and rituals of avoidance among New England fishermen. Mast 1 (1):66-78.
Pollack, S. 1989. Preparedness saved lives when dragger caught fire. National Fisherman 71 (December):13-15, 37.
Pontecorvo, G. 1986. Division of the spoils: hydrocarbons and living resources. Pp. 20-28 in The New Order of the Oceans: The Advent of a Managed Environment, G. Pontecorvo, ed. New York: Columbia University Press.
Providence Journal. 1990. 2 charged in sinking of boat, February 15. P. A-5.
Readings from . . . Alaska Seas and Coasts. 1979a. Fishing Vessel Safety—a checklist. Alaska Sea Grant Report 79-5. August. P. 155.
Readings from . . . Alaska Seas and Coasts. 1979b. Rigid polyurethane foam insulation & safety. Alaska Sea Grant Report 79-5. August. P. 167.
Rettig, R. B. 1986. Overview. Pp. 5-32 in Fishery Access Control Programs Worldwide: Proceedings of the Workshop on Management Options for the North Pacific Longline Fisheries, N. Mollet, ed., Alaska Sea Grant Report No. 86-4. Fairbanks, Alaska: University of Alaska.
Rice, D. P., E. J. MacKenzie, and Associates. 1989. Cost of Injury in the United States: A Report to Congress 1989. Atlanta: Centers for Disease Control. 282 pp.
Roberts, R. A. 1989. Fire on board. Westcoast Fisherman 4 (October):65-67.
Rostad, M. 1989a. Proper recognition for Peggy, the weather woman. Alaska Commercial Fisherman 1 (October):14-15.
Rostad, M. 1989b. Survival at sea: more than a sweet dong. The Fishermen's News (March):8-10.
Sabella, J., ed. 1986. North Pacific Fishing Vessel Owners' Association Vessel Safety Manual. Seattle, Wash.: North Pacific Fishing Vessel Owners' Association.
Sabella, J. 1987. Seattle's fishing vessel safety program. Alaska Marine Resource Quarterly 2 (First Quarter):14-15.

Sabella, J. 1989. The Cold Water Survival Handbook. Seattle, Wash.: John Sabella and Associates, Inc.

Safety at Sea. 1989. Protective clothing. (December):10, 12, 14-16.

Sainsbury, J. C. 1985. Vessel design. In Official Transcript Proceedings Before National Academy of Sciences, Marine Board, Symposium on Fishing Vessel Safety, Washington, D.C., December 11, 1985. Pp. 65-83.

Salit, R. 1989. Firm rejects claim on Andromeda. Gloucester Daily Times, February 17. Pp. A1, A3.

Shafer, S., and R. Beemer. 1984. Hypothermia. Reprinted from Pacific Fishing Magazine in Alaska Sea-Grams Number 11. December.

Shafer, T. 1990a. One day's crab fishing. In Attachment 2, Regional safety data: West Coast region, of the West Coast commercial fishing vessel safety assessment, prepared by R. Jacobson, G. Goblirsch, and F. Van Noy. Unpublished report to the Marine Board, National Research Council, Washington, D.C. Pp. A-1 to A-4.

Shafer, T. 1990b. One day's troll fishing. In Attachment 2, Regional safety data: West Coast region, of the West Coast commercial fishing vessel safety assessment, prepared by R. Jacobson, G. Goblirsch, and F. Van Noy. Unpublished report to the Marine Board, National Research Council, Washington, D.C. Pp. A-5, A-6.

Siegel, R. A., and W. G. Gordon. 1989. Fishery resources, trade, and economic growth. Paper presented at the Annual Meeting of the American Fisheries Society, Anchorage, Alaska, September 1989.

Sitka Sentinel. 1987. Halibut fishery dangerous. May 6.

Snyder, R. M. 1973. In Oceanography, the Last Frontier, R. C. Vetter, ed. New York: Basic Books.

Steiner, R. 1987. Sleep deprivation: surviving without sleep. Alaska Marine Resource Quarterly 2 (First Quarter):8-9.

Steinman, A. M., and P. S. Kubilis. 1986. Survival at Sea: The Effects of Protective Clothing and Survivor Location on Core and Skin Temperatures. Final Report No. CG-D-26-86, prepared for the U.S. Coast Guard. Washington, D.C.: Department of Transportation.

Stimpson, D. R. 1990. Mackerel fishermen square off over management plan. National Fisherman 71 (September):20-22, 36.

Stoop, J. 1989. Safety and bridge design. Paper presented at International Symposium: Safety and Working Conditions Aboard Fishing Vessels, L'Université du Québec à Rimouski, Quebec, Canada, August 22-24, 1989.

Sullivan, K. 1984a. City's fishing fleet facing insurance crisis. Gloucester Daily Times, July 16. Pp. A1, A4.

Sullivan, K. 1984b. 2 crews survive ordeals. Gloucester Daily Times, July 16. P. A5.

Sullivan, K. 1984c. Many say some boats are sunk intentionally. Gloucester Daily Times, July 19. Pp. A1, A9.

Sullivan, K. 1984d. Insurance nightmare in Gloucester: who's to blame? National Fisherman 65 (October):2-3.

Sutinen, J. G. 1986. Enforcement Issues in the Oyster Fishery of Chesapeake Bay. Paper presented to the Conference on Economics of Chesapeake Bay Management, Annapolis, Md. May 28-29, 1986.

Sutinen, J. G., and T. M. Hennessey. 1986. Enforcement: the neglected element in fishery management. In Natural Resources Economics and Policy Applications, E. Miles, R. Pealy, and R. Stokes, eds. Seattle: University of Washington Press.

Sutinen, J. G., A. Rieser, and J. R. Gauvin. 1989a. Compliance and Enforcement in Northeast Fisheries. Report prepared for the New England Fishery Management Council. May (Revised July 1989).

Sutinen, J. G., A. Rieser, and J. R. Gauvin. 1989b. Measuring and explaining noncompliance in federally managed fisheries. Unpublished paper. August (Revised July 1990).

Taylor, M. P. 1985. Underwriting and loss control. Pp. 15-25 in Official Transcript Proceedings Before National Academy of Sciences, Marine Board, Symposium on Fishing Vessel Safety, Washington, D.C., December 11, 1985.

Texas Shoreline. 1989. Hypothermia an ever-present danger in Gulf waters. 3 (December):1, 7.

Thompson, T. 1990. Safety on a shrimping vessel. Pp. A-7 to A-12 in Attachment 2, Regional safety data: West Coast region, of the West Coast commercial fishing vessel safety assessment prepared by R. Jacobson, G. Goblirsch, and F. Van Noy. Unpublished report to the Marine Board, National Research Council, Washington, D.C.

United States Code Annotated. 1990. Title 46, Shipping [Partial Revision]. St. Paul, Minn.: West Publishing.

U.S. Coast Guard. Undated, circa 1990. Port access control system: function requirements. Unpublished Coast Guard port security requirements planning document. Washington, D.C.: Department of Transportation.

U.S. Coast Guard. 1971. A Cost-Benefit Analysis of Alternative Safety Programs for U.S. Commercial Fishing Vessels, Volume I—Study. Washington, D.C.: Department of Transportation.

U.S. Coast Guard. 1982. Navigation and Vessel Inspection Circular No. 10-82 (Change 2, September 18, 1989). Washington, D.C.: Department of Transportation.

U.S. Coast Guard. 1983. Fish processor and cannery tender losses. June 29, 1983 (Revised September 8, 1983). Unpublished. Office of Merchant Marine Safety. Washington, D.C.: Department of Transportation.

U.S. Coast Guard. 1985. Commandant Instruction 16000.7: Marine Safety Manual, Volume II. Washington, D.C.: Department of Transportation.

U.S. Coast Guard. 1986a. Commandant Instruction M16247.1: Maritime Law Enforcement Manual (LEMAN). Washington, D.C.: Department of Transportation.

U.S. Coast Guard. 1986b. Navigation and Vessel Inspection Circular No. 5-86. Washington, D.C.: Department of Transportation.

U.S. Coast Guard. 1986c. First Coast Guard District Instruction M16703.1: Fisherman's Digest, issued April 1, 1986. Boston, Mass: Department of Transportation.

U.S. Coast Guard. 1988a. Commandant Instruction M10470.10B: Coast Guard Rescue and Survival Systems Manual. Washington, D.C.: Department of Transportation.

U.S. Coast Guard. 1988b. Commanding Officer, USCG Marine Safety Office Juneau letter 3100, April 21, 1988.

U.S. Coast Guard. 1989a. Navigation and Vessel Inspection Circular No. 4-89. Washington, D.C.: Department of Transportation.

U.S. Coast Guard. 1989b. Commanding Officer, USCG Marine Safety Office Juneau letter 3100, June 23, 1989.

U.S. Coast Guard. 1990a. Commandant (G-MTH-3) letter serial 5760/DnVC and 16716/DnVC, July 13, 1990.

U.S. Coast Guard. 1990b. Commandant (G-MTH-3) letter serial 16716, July 27, 1990. Subject: Interim Enforcement Program for Fish Catcher-Processor Vessels.

U.S. Coast Guard. 1990c. Statistics of Casualties—1987. Proceedings of the Marine Safety Council 47 (January-February):23-30.

U.S. Congress, House of Representatives. 1985. Fishing Vessel Insurance—Part 1. Serial No. 99-28. Washington, D.C.: Government Printing Office.

U.S. Congress, House of Representatives. 1987. Fishing Vessel Compensation and Safety. Serial No. 100-21. Washington, D.C.: Government Printing Office.

U.S. Congress, Senate. 1985. Fishing Vessel Safety and Insurance. S. Hrg. 99-268. Washington, D.C.: Government Printing Office.

U.S. Congress, Senate. 1987. Commercial Fishing Industry Vessel Safety and Compensation Act of 1987. S. Hrg. 100-346. Washington, D.C.: Government Printing Office.

U.S. General Accounting Office. 1990. Coast Guard: Strategic Focus Needed to Improve Information Resources Management. Report to the Chairman, Committee on Merchant Marine and Fisheries, House of Representatives. Washington, D.C.

U.S. Maritime Administration. 1979. Marine Fire Prevention, Firefighting and Fire Safety. Washington, D.C.: Government Printing Office.

Waage, N. 1990. Winter wind storm pounds the islands. Kodiak Daily Mirror, March 9. Pp. 1, 10.

Walker, M., and D. Lodge. 1987. Via the classroom: reducing vessel casualties. Alaska Marine Resource Quarterly 2 (First Quarter):17-18.

Walker, S. 1990a. Capsize. National Fisherman 71 (August):29-31.

Walker, S. 1990b. Giant halibut is a payback for a slow season. National Fisherman 70 (April):27-29.

Walter, C., and W. J. Morani, Jr. 1986. The sinking of the F/V Atlantic Mist. Proceedings of the Marine Safety Council (October):228-231.

Westcoast Fisherman. 1989a. EPIRBs: an update on the availability and the coming regulations. 4 (October):61-63.

Westcoast Fisherman. 1989b. News briefs: daily fisheries bulletin back on the air. 4 (October):12.

Wiese, C. 1978. Vessel tonnage: what it does and doesn't mean. (Originally printed in Alaska Seas and Coasts, October 1978). In Readings from . . . Alaska Seas and Coasts. Alaska Sea Grant Report 79-5. August 1979.

Wiese, C., Panel Moderator. 1988. Broker's views on insuring the fishing industry. Pp. 33-52 in Summary Proceedings of the National Workshop on Fishing Vessel Insurance and Safety, B. R. Melteff, ed., Alaska Sea Grant Report No. 88-2, Fairbanks, Alaska.

Wiese, C. 1990. Overcapitalization in Alaska groundfish fisheries: are we there yet? Bulletin of the School of Fisheries and Ocean Sciences 3 (June):7-8. University of Alaska, Fairbanks.

Williams, T. 1990. Incite: the exclusion of sea turtles. Audubon (January):24-33.

Yoder, E. 1990. For those in peril on the sea: progress in commercial fishing safety. Nor'Easter (Spring):9-13.

Zanoni, M. M. 1988. "I love you both. Goodbye." Proceedings of the Marine Safety Council (August-September). Pp. 143-145.

Zamiar, G. 1982. Fire in the hold. National Fisherman Yearbook 1982. P. 62.

Zapata Haynie Corporation. 1989. Vessel Operations Manual. Hammond, La.

Zimmerman, P. C. 1989. Tragedy of the F/V Vestfjord: wrong place at the wrong time. Pacific Fishing (August):29-33.

APPENDIXES

APPENDIX A

Committee Member Biographies

ALLEN E. SCHUMACHER, *Chairman*, worked in the field of marine insurance (particularly in relation to large, oceangoing commercial vessels) for over 37 years before retiring in 1987. During his career, he served in various claims, underwriting, and management capacities, including positions as chairman of the board and chief executive officer of the American Hull Insurance Syndicate (1971-1986) and chairman of the United States Salvage Association, Inc. (1971-1987). Mr. Schumacher also served on the Board of Managers (18 years) and Management Committee (10 years) of the American Bureau of Shipping, as chairman of the Maritime Transportation Research Board (1981-1982), and as vice-chairman of the Marine Board (1982-1983). He received a B.A. in economics from the University of California, Berkeley.

WILLIAM G. GORDON, *Vice-chairman*, is vice-president for programs and director, New Jersey Marine Sciences Consortium. Mr. Gordon retired from the National Oceanic and Atmospheric Administration (NOAA) in February 1987 as special assistant to the administrator, responsible for coordinating fisheries activities and programs with other federal agencies and foreign governments, and prior to that as NOAA's Assistant Administrator for Fisheries. He has held numerous other posts directed at strengthening scientific research capabilities, encouraging international programs, and directing fisheries management programs. He received a B.S. in zoology from Mount Union College and an M.S. in fisheries from the University of Michigan, where he continued with postgraduate studies.

BRUCE H. ADEE is associate professor of mechanical engineering at the University of Washington (UW) and served as director of its Ocean Engineering Program for over 10 years. He has been very active in many aspects of commercial fishing vessel safety and was instrumental in establishing the Fishing Vessel Safety Center at the UW. Dr. Adee is an expert in fishing vessel stability and has modeled the effects of various factors on stability. He is a member of the U.S. Coast Guard advisory committee on fishing vessel safety and an active member of the Society of Naval Architects and Marine Engineers. Dr. Adee received a B.S.E. from Princeton University and M.S.E. and Ph.D. degrees in naval architecture from the University of California, Berkeley.

DESMOND B. CONNOLLY is president of Independent Marine Services, Inc., where he surveys commercial vessels for condition and value, damage, and new construction and performs audio gauging. Prior experience includes 20 years in the U.S. Coast Guard (USCG), 11 years of sea duty, and 9 years with Marine Inspection offices. Mr. Connolly is a USCG Master of ocean steam or motor vessels (1,600 gross tons), USCG Second Mate of ocean steam or motor vessels (any gross tonnage), and Radar Observer and Adjustor of ocean and small craft. He is active in numerous professional maritime organizations and is the technical vice-president of the National Association of Marine Surveyors, Inc.

JOHN E. deCARTERET is an independent marine safety consultant, with 30 years of experience in marine vessel safety from his service in the Coast Guard. He served in every capacity of vessel operation and inspection, from deck watch officer to Chief of the Marine Safety Division, 13th Coast Guard District. Mr. deCarteret is well known as a marine safety expert emphasizing casualty and personnel investigations, surveys and inspections, litigation testimony, regulatory compliance, and pollution prevention and abatement. Mr. deCarteret received a B.S. degree from the U.S. Coast Guard Academy and pursued advanced studies at Seattle Pacific University.

GUNNAR P. KNAPP is associate professor of economics in the Institute of Social and Economic Research and the School of Public Affairs at the University of Alaska, Anchorage. Dr. Knapp's area of expertise is resource economics. He has studied the relationship between fisheries management and fishing vessel safety and has an extensive list of publications relating to fishing vessel accidents. Dr. Knapp received B.A. and Ph.D. degrees (economics) from Yale University.

HAL R. LUCAS is safety and loss-control manager for Sahlman Seafoods, Inc., where he is responsible for the safety program for a fleet of 100 U.S. registry shrimp boats. He developed shrimp boat safety self-inspection for the Southeastern Fisheries Association membership and designed a computer program for accident analysis, frequency, and severity measurements. His experience includes consulting on all phases of commercial insurance loss-control functions. Mr. Lucas received preengineering training at the University of Florida and is a graduate of the Supervisory Management Program at the University of South Florida and the U.S. Army Ordinance School.

JAMES O. PIERCE II is professor and department chairman of safety science, and director of the National Institute for Occupational Safety and Health, Region IX Southern California Educational Research Center, at the Institute of Safety and Systems Management of the University of Southern California. An internationally known expert in occupational safety and health, he has over 26 years of experience in graduate education, research, and administration at major universities. Dr. Pierce also served as special assistant for health technology with the Occupational Safety and Health Administration. He is active in various National Institutes of Health study sections. Dr. Pierce received a B.S. (chemistry) from the University of Alabama and M.S. and Sc.D. degrees (industrial health) from the University of Cincinnati.

LARRY D. SUND is vice-president in charge of operations for Golden Age Fisheries, where his management responsibilities include four factory trawlers and a joint venture/wet fish trawler. His prior positions, as vice-president and president of Jensen Maritime Consultants, Naval Architects and Marine Engineers, provided him with extensive design experience on at least nine different types of ocean fishing vessels. Mr. Sund has over 20 years of experience in the maritime industry, including a number of years as a commercial fisherman. Mr. Sund received a B.S. in mechanical engineering from the University of Washington.

BRIAN E. TURNBAUGH, owner and captain of a dragger, has been a fisherman and vessel owner for over 30 years in groundfish, shellfish, finfish, and lobster fisheries. He serves on the Board of Directors of the Point Club, a mutual self-insurance group, and chaired the club's committee that developed technical survey underwriting criteria; is a member of the Rhode Island Finfish Committee; and served on the Board of Directors of the Point Judith Fishermen's Cooperative. He has also worked as a consultant conducting market research for Maine shrimp, as field coordinator for a study of thermal effluents on the Hudson River, and as project leader for an environmental study of New Haven Harbor. He received a B.S. degree from the University of Maine.

JACK R. WILLIS is manager of safety, training, and security for Zapata Haynie Corporation, the largest commercial fishing company in the United States. He is familiar with all aspects of safety preparation and management, including licensing of crew members (all Zapata Haynie vessels are over 200 gross tons), preseason training, vessel inspection, and drug prevention and control programs. He also has extensive practical experience from sea duty in the U.S. Navy (20 years of experience on a destroyer, a minesweeper, and three diesel submarines) and master and mate sailing experience in seagoing tugs and passenger vessels. Mr. Willis received a B.A. from the University of Kansas.

MADELYN YERDEN-WALKER was formerly assistant training administrator with the State of Alaska Department of Education, Alaska Vocational Technical Center. She was directly responsible for schoolwide instructional program and facilities and was instrumental in the development of a fisheries and nautical training program emphasizing experiential training in marine safety, fisheries, gear technology, license preparation, and marine refrigeration and electronics. Ms. Yerden-Walker is active in various professional societies, including the Alaska Marine Safety Educators, Alaska Council of Local Administrators, and the Women's Fisheries Network. She received B.S. (biology) and M.Ed. degrees from Idaho State University.

APPENDIX

B

Papers Prepared for This Study

Bourke, R. 1990. Regional assessment for Hawaii and distant waters fisheries region. Unpublished contract report to the Marine Board, National Research Council, Washington, D.C.

Gordon, W. G. 1989. U.S. commercial fisheries: A brief review. Unpublished contract report to the Marine Board, National Research Council, Washington, D.C.

Hollin, D. 1990. Gulf of Mexico regional assessment. Unpublished contract report to the Marine Board, National Research Council, Washington, D.C.

Jacobson, R., G. Goblirsch, and F. Van Noy. 1990. West Coast commercial fishing vessels safety assessment. Unpublished contract report to the Marine Board, National Research Council, Washington, D.C.

Jones, R. 1990. Regional assessment for Caribbean region. Unpublished contract report to the Marine Board, National Research Council, Washington, D.C.

Jones, R., and J. Maiolo. 1990. South Atlantic regional assessment. Unpublished contract report to the Marine Board, National Research Council, Washington, D.C.

Munro, N. 1990. Regional assessment for Alaska region. Unpublished contract report to the Marine Board, National Research Council, Washington, D.C.

Nixon, D. W. 1990. Regional assessment for the Northeast region. Unpublished contract report to the Marine Board, National Research Council, Washington, D.C.

APPENDIX

C
International Fishing Industry Safety Activities

Although the U.S. commercial fishing industry fleet has not been heavily regulated for safety, the United States has participated in international forums addressing engineering and technical aspects of fishing vessel safety. These forums have led to development of extensive design, construction, and safety guidelines pertaining to vessels and corresponding safety and training guidelines pertaining to personnel. However, actual safety-improvement initiatives have been significantly less progressive in the United States than in other industrialized fishing nations.

Safety programs implemented by fishing nations have focused on vessel quality, operator licensing, manning standards, and compulsory safety training. These programs vary. Canada, Norway, and the United Kingdom, for example, have extensive requirements, while other countries are less stringent. Generally, vessels about 50 feet (15 meters) or larger are addressed; however, some countries address vessels as small as 30 feet (9 meters), such as New Zealand, and 40 feet (12 meters), such as the United Kingdom. Canada is considering extending vessel regulations to small fishing vessels, and some countries are beginning to focus more effort on practical training (T. Staalstrom, Det norske Veritas, personal communication, 1990; IMO News, 1990; Bárdarson, 1984).

This appendix provides an overview of international fishing vessel safety activities. Because of the similarities between Canadian and northern U.S. commercial fishing, Canadian activities are described in more detail. Information concerning fishing industry licensing programs administered by fishing nations was obtained by the U.S. Coast Guard and made available for this study. It was supplemented with material on licensing and vessel inspection provided directly to the committee, interviews with Canadian Coast Guard (CCG) officials, and published and unpublished literature. The material, while incomplete, provides general insight into approaches taken by other fishing nations.

INTERNATIONAL MARITIME ORGANIZATION

The International Maritime Organization (IMO, formerly the International Maritime Consultative Organization, or IMCO, from 1960 to 1982), is the United Nations marine safety organization. The IMO has been working to improve the safety of commercial fishing vessels for years. The U.S. Coast Guard has actively supported this work since about 1966. At that time, the Coast Guard supported an IMCO panel on fishing vessel stability. Since 1968, the IMCO subcommittee has been the only subcommittee solely dedicated to an individual class of vessel. The Coast Guard is designated by the U.S. State Department as the federal marine safety agency and is empowered to represent U.S. marine safety and environmental interests to the IMO.

United States involvement in IMCO fishing vessel safety work was a cause for concern, since there was no enabling authority for any U.S. agency to promulgate regulations pertaining to fishing vessel design or construction, the method by which international conventions are brought into force by member nations. The Coast Guard nevertheless approached the State Department with a plan to voluntarily support international fishing vessel safety activity. The agency believed this would be a way to promote essential research leading to operation, design, construction, stability, and safety guidelines that would benefit U.S. fishermen. The State Department approved this approach.

During the mid-1960s, an international fishing vessel safety program was started by three United Nations agencies—the International Labor Organization (ILO), the Food and Agriculture Organization (FAO), and IMCO (Appave, 1989; Plaza, 1989). This activity led to a two-part code of safety for fishermen and vessels:

- Part A, operational safety guidelines (printed in the mid-1960s), and
- Part B, design and construction aspects of fishing vessels, which was purely voluntary guidance.

Initial IMCO input to Part B began as specialized information on stability of fishing vessels. This was subsequently broadened to cover the entire design and construction of fishing vessels, including fire safety, engine machinery safety, radio communications safety, and all other safety functions. Work on Part B progressed from 1968 to 1974.

In 1974, the IMCO Assembly scheduled an international convention for 1977 on the safety of fishing vessels. Three years were spent preparing for the conference, which culminated in the Torremolinos International Convention for the Safety of Fishing Vessels, 1977. The United States sent a full delegation, led by a Coast Guard flag officer, to the conference. Since 1977, U.S. delegations have remained active and have considered extending safety guidelines to smaller fishing vessels in a new code discussed by the IMO Subcommittee on Stability, Load Lines, and Safety of Fishing Vessels, which meets about once a year.

The Torremolinos International Convention has not yet entered into force, principally as a result of technical requirements unacceptable to some nations. The IMO is developing a protocol to the convention to resolve these impediments, but recognizes that more work is needed internationally to improve commercial fishing vessel safety, with consideration given to vessel standards and training of personnel (IMO News, 1989, 1990; Plaza, 1989).

The Coast Guard's international work, even in the absence of enabling authority for regulations, resulted in a flow of information to the U.S. commercial fishing industry through the Coast Guard and the National Marine Fisheries Service. The Coast Guard has routinely invited naval architects and fishing industry representatives to participate on its advisory committee for the U.S. IMO subcommittee delegation. As a result, Coast Guard involvement has provided the technical insight needed to develop its voluntary guidelines (consolidated in NVIC 5-86) and to support development of fishing vessel safety manuals for the North Pacific, Gulf, and Atlantic coasts.

Canadian fishing vessel safety regulations and activities are described in the section that follows, and those of other representative fishing nations are summarized in Table C-1.

CANADIAN VESSEL SAFETY RULES AND REGULATIONS

The Canadian commercial fishing industry is similar in some respects to the industry in northern U.S. waters, for example, gear types, stocks exploited, and environmental conditions. Commercial fisheries are a prominent element of the Canadian economy and are a major industry in coastal communities. Considerable provincial and national attention is focused on the fishing fleets, and national concern for continuing high fatalities and vessel losses prompted a CCG fishing vessel safety study. Analysis of fishing vessel search and rescue (SAR) cases in Canadian waters strongly implicated human error as a primary cause of vessel losses and fatalities. The CCG also found that a large portion of casualties was associated with smaller fishing vessels (CCG, 1987).

Between 1982 and 1986, the primary causes of loss of life in the Canadian fishing industry were man overboard, capsizing, foundering (sinking), grounding, fire, and explosion. High-value fisheries, which induce increased risk taking for profit, had a disproportionate share of the fatalities. Fisheries implicated were seining, shellfish (lobster and crab), and groundfish, especially where small fishing vessels operated at considerable distances offshore. About 90 percent of the fatalities involved vessels under 60 gross tons, and 60 percent involved vessels under 15 gross tons. The primary causes of vessel losses were fire and

TABLE C-1 Synopsis of Representative International Fishing Vessel Safety Rules and Regulations

Regulation and Infrastructure	Registration and Licensing	Education and Training
AUSTRALIA		
Commonwealth regulates the safety of fishing vessels engaged in overseas operations. Uniform codes (Australian Transport Advisory Council) apply to 5 classes of fishing vessels categorized by operating area: unlimited operations, offshore operations, restricted offshore operations, sheltered water operations for partially smooth of smooth water, and sheltered water operations for smooth water only.	Extensive licensing requirements for all classes of vessels. Qualifications required for skippers, mates, engineers, and crew.	Extensive training requirements for all classes of fishing vessels. Crew members on overseas vessels must undergo basic safety and survival training.
DENMARK		
Danish Maritime Authority operates all training ships and maritime training institutions. Education is coordinated with the maritime industry through the Maritime Education Council, an advisory group representing seafarers and fishermen's associations.	Masters must be Danish citizens and hold a 1st or 3rd Class Skipper's Certificate of Competency, which certifies criteria have been met to operate certain size fishing vessels on limited or open waters.	Three general phases for training are: (1) noncompulsory pre-sea training--5 weeks of entry level, general knowledge, on shore training for those with at least 9 years of schooling; and skipper school training with (2) 3rd class, and (3) 1st class programs. Skipper school requires 24 months seagoing service, 12 of which must be aboard fishing vessels. Training for 3rd and 1st class examinations is compulsory.

TABLE C-1 (*continued*)

Regulation and Infrastructure	Registration and Licensing	Education and Training
FINLAND		
Vessels 8.5-27.5 m engaged in open-sea fishing in the Baltic Sea and contiguous gulfs are regulated by the National Board of Navigation. Those > 27.5 m or fishing outside of specified Baltic waters are regulated under merchant vessel enabling authorities. Criteria for masters and machine attendants on board fishing vessels include medical fitness and certification.	Fishing vessels must have master with Skipper's Certificate A and B, and a machine attendant with a Machine Attendant's Certificate A or B. Candidates for Skippers Certificates must prove their identity, be medically fit, and have proof of formal training.	Candidates for a Skipper's A or B Certificate need certificate from a maritime school or inspector covering specific expertise and serve 2 years on a vessel. Machine Attendant candidate B must serve 9 months in a motor vessel engineering department and have technical school or machine surveyor certificate attesting to his or her knowledge. Certificate A criteria expand in-service requirements.
JAPAN		
The government administers a comprehensive safety program for all commercial fishing vessels, which includes compulsory licensing and manning standards. Mandatory safety inspections are required for all Japanese fishing vessels. Certain survey and technical standards are not applied to small fishing vessels used within 12 miles of the coast.	Ministry of Transport administers a comprehensive, multi-tiered licensing program for masters, mates, and engineers. Manning standards depend on vessel employment and tonnage for deck officers, and power output for engineering officers. The licensing program promotes officer competency.	

NEW ZEALAND Fishing vessels > 12 m long must be surveyed to determine the vessel's category. Vessels categorized as Class X fishing vessels are regulated for fire fighting and lifesaving equipment and maintenance requirements. There are extensive training and certification requirements for skippers and mates on board fishing vessels.	A Commercial Launchmaster certificate is required to fish within 12 miles of shore in vessels < 15.26 m long. To operate a fishing vessel up to 100 miles offshore, the vessel operator must be certificated as a New Zealand Coastal Master.	Launchmaster applicants must have 18 months sea service and pass an examination covering various topics. Coastal master applicants require 3 years sea service and must pass an examination. Training and examination requirements for mates and skippers of deep sea fishing boats include 4 years sea service and examination for mates; 5 years sea service and examination for skippers.
NORWAY Commercial fishing vessels > 15 m long are heavily regulated for operation and material condition.	Compulsory, comprehensive licensing for masters, mates, "master fisherman," deck officers, and engineering officers.	Basic requirements for licenses include prerequisite deck or engineering sea-service and 1-3 years practical training. Since 1986, safety training is compulsory for all fishermen aboard any type and size of vessel, except those operating on inland fresh waters.
SWEDEN Fishing vessels with overall hull length at least 9 m are regulated. Competency, medical, manning, and watch keeping requirements are in effect for operators, mates, and engineers. They vary by operating area and vessel size. Theoretical training at an educational facility, practical experience, and a medical certificate are required.	Masters of 9-12 m vessels must obtain a Skipper's examination certificate; no sea service requirements are specified; skippers of 12-24 m vessels require a B certificate; officers in charge of a watch in extended coastal or North Sea areas must have a Skipper's examination certificate. Vessels > 24 m and < 500 gross tons must have Skipper B for sheltered trade	Skipper B candidates must have examination certificate and 36 months' deck service, 12 of them aboard merchant vessels. Skipper A candidates must have 36 months' deck service, 18 of them in extended coast or extended trade, including 12 months aboard a merchant vessel; must complete the Skipper A or Skipper First Class course.

TABLE C-1 (*continued*)

Regulation and Infrastructure	Registration and Licensing	Education and Training
SWEDEN (*continued*)	and Skipper A for extended coast and North Sea; mates must hold Skipper B certificate. There are three categories of engineers: on vessels < 405 KW, one crew member must have passed an engineer examination; 405-750 KW, one crew member must hold Engineer B certificate; 750-1,600 KW, one crew member must hold Engineer A certificate.	Training and marine engineering service requirements are similar to those for deck officers, but in engineering disciplines.
UNITED KINGDOM Comprehensive regulations have been in force since 1975. Surveys and certification of fishing vessels > 12 m are required; they apply to about 2,000 vessels. For vessels > 16.5 m, deck officers and engineers have comprehensive entry level professional training, certification, manning, and watch keeping requirements.	There are 5 categories of fishing vessels and 3 levels of competency certification (Class 1, 2, and 3) for operators and engineers. Minimum deck officer manning is keyed to a combination of fishing area and vessel size.	New entrants and working fishermen must complete a basic training course or skipper or mate courses with basic elements. This training requirement is being phased in over 4 years with seasoned fishermen receiving the most time to complete the requirements. Owners and operators must ensure that new entrants produce satisfactory training course attendance certificates. Examinations and sea service requirements vary by class of certificate.

explosion, foundering (sinking), and grounding: 90 percent of vessels lost were under 60 gross tons. The CCG, while noting that the vast majority of fishermen have operated safely and successfully during their fishing careers, nevertheless attributed the majority of incidents resulting in the loss of 824 vessels and 140 fatalities to human error (CCG, 1987). It found that principal considerations leading to human error were:

- risk taking for profit induced by increased economic pressures,
- vessel design and safety sacrificed for economic benefit, and
- normal practice of seamanship disregarded or not known.

Concern for the well-being of Canadian fishermen also prompted examination of occupational safety and health aboard fishing vessels. The committee considered Canadian studies on occupational safety and health (Gray, 1986, 1987a,b,c). The report of a tripartite committee representing industry, federal, and provincial views that was appointed by Labour Canada to recommend corrective action was also considered (Canada, Government of, 1988; Carter, 1989).

Examining the relationships among human factors, plant and equipment, and regulations, Gray (1987c) found that:

- There are competent and fully qualified fishermen in all regions, "but varying numbers of untrained or partially trained persons who are attempting to learn the business by the process of exposure and survival."
- Safety in the fishermen's workplace is primarily the result of effective application of knowledge, training, and awareness of the job and reasons for accidents.
- There is strong industry support for education and training as the primary alternative for improving safety, but a lack of confidence in regulations unless they are developed in consultation with affected fishermen.

Gray further identified three basic principles for correcting occupational safety and health problems in the Canadian fishing industry and recommended that the principles be addressed in the following order (Gray, 1987c):

- establish an educated or trained work force,
- provide safe vessels and equipment, and
- establish simple, readily enforceable regulations.

The Labour Canada committee in its 1988 report (Canada, Government of, 1988) found that emphasizing training and education was the best way to improve safety performance in the fishing industry. They were "not persuaded that volunteerism in the field of training will materially alter the existing reality. Whether as a result of the rugged individualism which typifies the industry or an apparent discomfort with the educational setting, there seems to be a natural reluctance on the part of fishermen to submit to a formal training process." The committee recommended a compulsory safety training program "as a prerequisite to obtaining an annual, personal, commercial fishing license." Also recommended were:

- a training curriculum that, insofar as practical, is based on existing courses and facilities with flexibility to accommodate varying regional needs,
- public awareness programs targeted to high-risk fisheries and specific fishing seasons,
- a formal material inspection program for fishing vessels under 15 gross tons, and
- a mandatory self-inspection checklist completed annually as a prerequisite for obtaining a vessel fishing license.

The actions called for by the Labour Canada committee have been addressed in part by ongoing CCG safety initiatives associated with its study of fishing vessel safety (Carter, 1989; CCG, 1987). Training standards for marine emergencies have been in place for more than 10 years and were updated in 1988 (CCG, 1988). Certification for masters, mates, and engineers is required on vessels in excess of 100 gross tons, and this will soon be reduced to 60 gross tons. The CCG has conducted a public awareness program to communicate risks to fishermen in selected locations. A new fisherman's handbook has recently been published, incorporating voluntary self-inspection checklists. The handbook was initially distributed by the Canadian Department of Fisheries and Oceans to all fishermen with their annual fisherman's license. Inspection regulations for small fishing vessels are being developed but are progressing slowly, partly because of socioeconomic factors in some sectors of the fishing industry. New inspection regulations for large fishing vessels are on hold pending completion of action on small-vessel regulations. Existing fishing vessel regulations require immersion suits on all fishing vessels exceeding 150 gross tons. Thermal deck suits are proposed on all fishing vessels under 150 gross tons, and these amendments are expected to be in place by the end of 1990. Additionally, life rafts will be required on all fishing vessels regardless of size or location after March 1991. These efforts have not reached the stage at which their effectiveness can be determined.

EFFECTIVENESS OF INTERNATIONAL AND FISHING NATION SAFETY MEASURES

A major question is whether compulsory programs to overcome or mitigate vessel-related casualties or human factors through engineering or technical solutions will work. There is little material on program effectiveness in other countries. Studies have been conducted in Norway, the Netherlands, the United Kingdom, and Spain, for example, but they have tended to focus on training, statistics, and causes of accidents rather than on performance of technical systems in relation to compulsory programs. It appears that fatalities have generally been reduced, while rates of incidence for injuries related to vessel casualties and workplace accidents appear unchanged. The lack of apparent change in injury rates may be related to working conditions and methods, vessel design, training deficiencies, and changes in the numbers of fishing vessels and fishermen (Carbajosa, 1989; Dahle and Weerasekera, 1989; Hoefnagel and Bouwman, 1989; Hopper and Dean, 1989; Stoop, 1989). The number of vessel casualties has varied. For example, in the United Kingdom, since safety rules were applied to all vessels over 12 meters during the mid-1980s, the number of losses of these vessels has been significantly reduced. However, losses of vessels under 12 meters have more than doubled, perhaps partly because of a large increase in the number of vessels under 12 meters, to which only lifesaving and fire-safety government regulations apply (Hopper and Dean, 1989). At least for the Scottish inshore fishing fleet, new designs have favored beamy vessels in order to satisfy length-based legislation, with suspected but as yet unproven adverse changes in seaworthiness (Macleod and MacFarlane, 1989).

APPENDIX

D

Data Sources and Issues

NATIONAL SAFETY DATA

Potential data sources for assessing safety in the fishing industry at the national level include the U.S. Coast Guard, the National Transportation Safety Board, National Marine Fisheries Service, Occupational Safety and Health Administration, and Marine Index Bureau.

U.S. Coast Guard Data

Coast Guard data are maintained on a program by program basis. Some programs have multiple data bases that were developed independently and are not compatible because of computer operating systems, software, data format, or data fields maintained (see U.S. General Accounting Office, 1990). The principal Coast Guard data sources used for this report follow.

Main Casualty (CASMAIN) Data

At present, the U.S. Coast Guard CASMAIN data base is the only source of detailed information available on fishing vessel casualties in the United States. The CASMAIN data base is a coded summary of incidents reported on Coast Guard Marine Casualty Reports. While the CASMAIN data do not cover all commercial fishing vessels or injuries, they provide sufficient information to demonstrate that there are significant safety concerns in commercial fishing, and are useful for identifying the nature and proximate causes of vessel and personnel casualties. However, data on causes of casualties are incomplete. In addition, the CASMAIN data do not include information on the fishery in which vessels were participating when casualties occurred.

The CASMAIN data base includes different information for vessel and personnel casualties (fatalities and injuries). Relatively more information about the vessel, location of the incident, and environmental conditions at the time is provided for vessel than for personnel casualties. In general, better data are available on safety problems for documented than for undocumented vessels.

Reporting of marine casualties is required by 46 U.S.C.A. §6101. A "Report of Marine Accident, Injury, or Death" (Form CG-2692) must be filed with Coast Guard marine safety offices for all marine incidents resulting in:

- accidental as well as intentional grounding, which creates a hazard to navigation, the environment, or the safety of the vessel;
- loss of main propulsion or primary steering, or reduction of the vessel's maneuvering capabilities;
- occurrences materially and adversely affecting the vessel's seaworthiness or fitness for service, such as fire, flooding, or damage to bilge pumping systems;
- loss of life;
- serious injury; or
- any occurrence resulting in property damage in excess of $25,000.

Specific reporting criteria are found in 46 CFR Part 4.

Depending on the seriousness of the incident, further investigations or formal inquiries may take place. Records of these events are on file for 3 years at marine safety offices, after which they are permanently filed with the Coast Guard's Marine Safety Evaluation Branch, Marine Investigations Division, in Washington, D.C. More than 1,000 reports are filed each year for incidents on commercial fishing industry vessels. Filed with some reports are more extensive investigation records, but the thoroughness is uneven. That there is a record on file does not mean the Coast Guard was actively involved in the incident, although in many cases Coast Guard forces may have responded.

Coast Guard officials believe that the CASMAIN data base includes most fishing vessel casualties resulting in major damage to and most fatalities occurring on documented vessels. However, less serious vessel casualty incidents, such as temporary grounding or propulsion loss, may never be reported to the Coast Guard and thus may not be included in the CASMAIN data base. Similarly, many injuries are not reported. In addition, more serious incidents may not be reported if they occur on small, undocumented vessels and the Coast Guard does not become involved in providing rescue services. Thus, the CASMAIN data base may significantly understate the extent of injuries and minor vessel casualties (those not resulting in major damage) for all vessels, as well as fatalities or major vessel casualties for undocumented vessels.

Some vessel casualty incidents not recorded in CASMAIN may be recorded in Coast Guard search and rescue (SAR) data. However, the CASMAIN and SAR data bases are not compatible with regard to computer operating systems, software, or common key data fields, making cross-comparison difficult. Other incidents might be reflected in the Coast Guard's recreational boating safety (RBS) data base, inasmuch as a number of smaller commercial fishing vessels are converted recreational vessel hulls. Again, the data bases are not compatible.

CASMAIN data are believed to be reasonably complete for fatalities associated with documented fishing vessels, although it is difficult to screen out deaths not directly related to commercial fishing activities, such as suicides. Fatality data are believed to be somewhat less complete for deaths associated with state-numbered vessels. The Coast Guard believes that about 90 percent of marine commercial fishing fatalities are recorded in CASMAIN. One reason is that except for the state of Alaska, which issues death certificates for state residents lost at sea, only the Coast Guard, in the absence of an individual's remains, will issue a letter of presumptive death needed by heirs to settle estates. Coast Guard field stations also maintain liaison with local authorities, learning of some fishing industry fatalities via that medium.

Although reports are required for serious injuries, CASMAIN data on personal injuries are poor. The data are incomplete and of limited utility. The Coast Guard does not have a formal estimate of how many reportable injuries are recorded. No data were developed that provide insight on what the actual percentage is. It is known from such sources as the Alaska Fishermen's Fund (discussed later in this appendix), however, that the incidence of personal injuries in the harvest sector exceeds that recorded in CASMAIN. Many injuries do not meet the reporting thresholds; those that do may simply not be reported.

CASMAIN cause data, while incomplete, are nevertheless the best source of cause data for fishing industry vessel casualties. CASMAIN data coding allows for up to seven cause categories, but in practice, supplemental data identifying contributing or underlying factors to casualties are entered for only about 30 percent of recorded cases. Even the primary cause code is recorded as "unknown" for 40 percent or more of all accidents. Thus, CASMAIN data do not provide for thorough analysis of how human error, vessel or equipment failure, and environmental factors interact in causing vessel or personnel casualties.

Search and Rescue (SAR) Data

The Coast Guard maintains data on each SAR case in which Coast Guard forces assisted. This assistance can include everything from communications services to extensive surface and air searches for missing vessels and personnel. Data are not available for situations where external assistance was required but the Coast Guard was not involved.

The data provide very useful information about the utilization of Coast

Guard forces for SAR and the results of each case in which they are involved. The data are recorded by Coast Guard field forces on a standard collection form and entered into the SAR data base administered by the Search and Rescue Division at Coast Guard Headquarters in Washington, D.C.

SAR data include much less information than CASMAIN data and are not integrated with the CASMAIN data base. SAR data include both documented and state-numbered vessels, although they cannot be isolated. SAR data fields are general with regard to cause and not entirely compatible with CASMAIN, as noted previously. The data fields are primarily oriented toward vessel-related information, although limited information is available on the number of fatalities. Individual vessels cannot be isolated in the data to determine if they have been the subject of multiple SAR incidents. The actual number of vessels lost is not recorded. Rather, the data record the value of property lost or saved and severity of the incident. From the severity and nature-of-incident codes, it is possible to approximate the number of vessels totally lost. These coding dimensions significantly limit the utility of the data for analytical purposes, although the severity codes do provide a general sense of how serious the incident was.

An electronic mapping and density plotting capability for displaying SAR data was available for this study and used to develop density plots of SAR cases in which personnel or property were in danger of being lost or were lost. Sample density plots are shown in Chapter 3.

SAR data record only one cause data field for human factors—personnel error. The Coast Guard's SAR data system manual does not provide criteria for using this category, leaving determinations to the discretion of the reporting source. The single-dimensional nature of SAR data precludes analyzing the interactive or interdependent relationship of human factors in the chain of events producing a SAR case to other cause categories, such as propulsion or hull failure and fuel exhaustion. Man-overboard incidents are not categorized, nor is the availability, use, or proper performance of lifesaving equipment.

Summary Enforcement Event Report (SEER)

The SEER data base is maintained by the Operational Law Enforcement Division of Coast Guard Headquarters. Its function to provide data support for maritime law enforcement activities is oriented toward drug and fisheries enforcement. It records the number of boardings of commercial fishing industry vessels by Coast Guard forces. The only year for which complete data are available is fiscal year 1989. Documented and state-registered fishing vessels cannot be isolated in the data, although the total number of fishing vessel boardings is recorded. Data fields do not permit correlation with other Coast Guard data bases. Multiple boardings that occurred cannot be determined using existing data base software, but can be ascertained from an entry-by-entry review of each data file. SEER data also include violations of marine pollution, and safety and survival equipment laws and regulations. These are all coded as "boating safety" violations and cannot be separated for analysis.

The quality of individual boardings and the resulting documentation depend on the experience and expertise of boarding officers. Thus, some violations reported by boarding officers—estimates range from 5 to 15 percent—are screened out as technical nonviolations of pollution and uninspected vessel regulations prior to submitting violation reports for civil penalty action. As a result, SEER data have limited utility for safety analysis, though they proved useful for assessing the scope of underway boarding activity affecting commercial fishing industry vessels.

Marine Safety Information System (MSIS)

MSIS is a proprietary, on-line, computer-based electronic information system for marine safety information. Information is recorded by vessel. Among the data maintained are boarding histories of commercial vessels. There is no direct capability to interrogate MSIS for aggregate data, type of fishing industry vessel, or utilization within the industry. The information in MSIS is compatible with CASMAIN and can be downloaded. In this manner, some aggregation of current information may be possible, but was beyond the scope of this study.

The Coast Guard has initiated a project (Marine Safety Management System) to improve cross-referencing of all marine safety data collected by the Service, including that maintained in MSIS, CASMAIN, and other data bases maintained by the Office of Marine Safety, Security and Environmental Protection in Washington, D.C. The Coast Guard is also developing a prototype Marine Safety Network (MSN) as a technologically advanced replacement for MSIS.

Recreational Boating Safety Data

Fishing vessels are eligible for courtesy marine examinations (CMEs) offered by the Coast Guard Auxiliary in support of recreational boating safety. Accidents involving recreational vessels are required to be reported by law. The data are recorded in a data base maintained by the Auxiliary, Boating and Consumer Affairs Division at Coast Guard Headquarters. The data collection form (CG-3865) includes data fields for "commercial activity" and "fishing." However, the Coast Guard's boating safety data base is not used to discriminate between recreational and uninspected commercial fishing vessels for either CMEs or boating accident analysis. This precluded an assessment of accidents involving state-numbered commercial fishing vessels for which boating accident reports may have been filed. Furthermore, Auxiliary activity is almost exclusively oriented toward the recreational public. Uninspected fishing vessel use of the CME public service is considered negligible.

National Transportation Safety Board Data

The National Transportation Safety Board (NTSB) maintains a data base of fishing industry accidents it considers. The data are limited to a small number of major casualties. NTSB reports of these casualties reflect safety problems suspected to be endemic to the industry, but the data base does not provide the means for determining the safety records of vessels industrywide.

National Marine Fisheries Service Data

The National Marine Fisheries Service (NMFS) conducts extensive compliance boardings of uninspected fishing vessels under fisheries management regulations. However, NMFS officials do not check for safety or survival equipment. The NMFS Enforcement Management Information System contains no information relevant to vessel safety. However, estimates of landings and aggregate numbers of vessels active in commercial fishing are collected from regional sources and maintained in a national data base. While not complete, this information can be used to provide a general indication of fleet size. NMFS regional data sources are discussed later in this appendix.

Occupational Safety and Health Administration Data

Both the Coast Guard and the Occupational Safety and Health Administration (OSHA) have responsibilities in this area. OSHA has enabling authority to regulate occupational safety and health aboard certain uninspected fishing industry vessels that are not inspected by the Coast Guard. So far, the Coast Guard has not preempted OSHA with regard to regulation of uninspected fishing vessels (see Expert, 1989), and OSHA's involvement has been limited to regulating industrial activities aboard fish processors employing more than 10 workers.

OSHA compliance inspections are conducted periodically and the results recorded in a data base. OSHA data base files were reviewed by the committee. They contain limited information about processing-line accidents, but negligible information on operational safety.

The Coast Guard has sponsored extensive research concerning marine occupational safety, principally focused on hazardous materials. Standards in this area are being considered. The work to date may have some application aboard large processing vessels, but appears to have limited application for the majority of fishing industry vessels.

Marine Index Bureau Data

The Commercial Fishing Claims Register (CFCR)

Insurance companies keep records of claims filed by commercial fishing vessel owners resulting from both vessel and personnel casualties. However, these claims data have not generally been available to researchers. Moreover, claims submitted to any particular insurance company are not necessarily representative of the types of incidents occurring in the fishing industry as a whole. However, not all companies have participated in sending this information to the CFCR, and since even these participating companies do not send records of all claims, the CFCR data cannot be considered a representative sample of injuries occurring in the commercial fishing industry.

The collection form employed for this data base contains data fields that could be effectively cross-tabulated for both personal injury and hull and machinery data. However, in practice, only the injury sections of the CFCR forms have been completed by voluntary contributors from the marine insurance industry. The injury data are very incomplete and are principally from the North Atlantic and Pacific Northwest, and the population of insured vessels represented is not known. CFCR data are not sufficiently complete to permit meaningful analysis, but could prove very valuable if they were more fully developed.

Vital Statistics Data

Death certificates provide an alternative source of data on commercial fishing fatalities. As noted in Chapter 3, proportional mortality rate (PMR) analysis can be used to track occupational fatalities in commercial fishing, even in the absence of population-at-risk and employment data.

Death certificate data may also be used to validate Coast Guard fatality data. As part of its West Coast regional assessment, death certificate data for Washington State were compared with Coast Guard CASMAIN records. This disclosed that CASMAIN fatality data accounted for nearly all deaths recorded for Washington State commercial fishermen. However, there are significant limitations to death certificate data as an immediate solution to data problems in assessing safety in the commercial fishing industry. Although death certificates in most states provide information on the deceased's occupation, in many states it is not computer coded. In addition, it is usually not possible to discern part-time from full-time commercial fishermen or to obtain comprehensive information on circumstances surrounding the fatality.

Published vital statistics are frequently organized by occupational groups using common denominators meaningful to the public at large. The common frames of reference are often deaths per thousand or hundred thousand. Similar denominator-based approaches are used to convey comparative information about vessel casualties. Applying analytical technique to safety assessments in the commercial fishing industry had very limited utility for this study. Where occupational mortality was recorded in national data bases, it was aggregated with agriculture and forestry statistics. Furthermore, unless vastly improved data become available, particularly with regard to populations at risk and exposure variables, denominator-based statistics to assess safety problems or monitor safety performance must be used with great care. Proportional statistics hold promise for monitoring changes in the fishing industry, but have been applied in only one state.

Plans for Development of New Data Sources

Recognizing the lack of data on commercial fishing occupational illnesses and injuries, the Commercial Fishing Industry Vessel Safety Act of 1988 (CFIVSA) included provisions calling for the development of better data. Proposed regulations released by the Coast Guard would require the owner or individual in charge of the vessel to report any "injury to an individual that causes that individual to remain incapacitated for a period in excess of 72 hours" either to the insurance underwriter for the vessel or to the Marine Index Bureau. The underwriter of primary insurance would have to report each casualty to the Marine Index Bureau within 90 days of being notified and when it paid a claim. It is not yet known when this reporting system may be instituted and to what extent the information will be available for research on occupational illnesses and injuries in the commercial fishing industry.

REGIONAL SAFETY DATA

Regional Assessments

A large volume of data was assembled for this study. The data available at the national level were not complete and in most cases could not be effectively correlated. As a result, regional assessments were commissioned to supplement national data on the numbers and status of uninspected fishing industry vessels and the population at risk. The regional data varied greatly in availability as well. Generally, the more reliable data were available for the West Coast and Alaska. But, considerable insight for each region was obtained through the process. Proportional mortality rates were available for fishermen domiciled in Washington State. Injury data were very limited except for Alaska; some injuries were compensated for through the Alaska Fishermen's Fund, which provides a partial injury data resource. Inadequacies with virtually all the data made it difficult to normalize them to develop casualty rates. The following sections provide a synopsis of major regional data sources.

North Atlantic Region

The NMFS Northeast Fisheries Science Center, Woods Hole, Massachusetts, maintains data on catch effort and operating units in the fisheries. The data are based on an interview system in which NMFS statistical agents are assigned to major fishing ports throughout the Northeast and mid-Atlantic states to collect data at landing sites. No federal or state records documenting actual fish landings were identified. Fish auction records are available concerning estimated catches.

South Atlantic and Gulf/Caribbean Regions

The NMFS Southeast Regional Fisheries Science Center, Miami, Florida, administers an interview system similar to that found in the North Atlantic region. The system is less robust because of the size of the regions and dispersion of the fishing fleet. Formal documentation of actual fish landings is not maintained at either the federal or state levels.

West Coast Region

PACFIN Research Data Base

All fish landings at West Coast and Alaskan ports are formally documented by a "fish ticket" for every vessel that lands fish commercially on the West Coast. These data are collected by the states and include the number of landings, specific data on the vessel's state of registry, port operated from, vessel information—including whether documented or state-numbered—species landed, and date. The data are provided by the states to the NMFS Southwest Fisheries Science Center, La Jolla, California, and the NMFS Northwest and Alaska Fisheries Science Centers in Seattle, Washington, and archived in the PACFIN research data base in Seattle.

The PACFIN data were available to the committee on a very limited basis. They are excellent for monitoring fishing activity, but do not provide exposure data. It may be possible to estimate exposure data based on the data and local knowledge about fishing practices.

Alaskan Region

The fish ticket information discussed above applies, as does the discussion in Chapter 3 of how this information and vessel and personnel licensing information maintained by the Commercial Fisheries Entry Commission in Juneau, Alaska, could be more fully exploited for Alaskan fisheries.

A source of data on fishing injuries in Alaska is claims to the Alaska Fishermen's Fund, which was established in 1951 to provide for treatment and care of Alaska licensed commercial fishermen who are injured or become disabled while engaged in commercial fishing in Alaska. All persons holding Alaska commercial fishing licenses are eligible for up to $2,500 for medical expenses. In recent years, the Fishermen's Fund has required fishermen to make claims first on other insurance, if available.

Data are recorded and coded for date of incident, nature and location of injury, location of fishing grounds, type of gear, and dollar value of claim. Data are recorded but not coded for fishery, are currently not maintained in an automated data base, and are not routinely published.

Even though several thousand claims are submitted each year, the Fishermen's Fund does not provide a complete record of injuries occurring in the Alaska fishing industry, since its use is voluntary and, according to the fund administrator, many fishermen use other sources to pay for medical claims. Nevertheless, it provides useful information about injuries occurring in the Alaska commercial fishing industry.

Hawaii/Southwest Pacific Region

Fish landing data for this region are included in the PACFIN data but are based on an interview system and the port of offloading. The data are collected by the states and territories and provided to NMFS.

APPENDIX E
Selected CASMAIN Vessel Casualty and Fatality Data

Tables follow on next pages

TABLE E-1 Documented Fishing Industry Vessel Casualties, 1982-1987, by Primary Nature of Casualty

Primary Nature of Casualty		Number of Vessel Casualties	Total Vessel Losses	Total Vessel Damages (dollars)	Number of Vessel-Related Fatalities
CAPSIZING		**138**	**94**	**32,033,117**	**114**
COLLISION		**739**	**133**	**28,397,240**	**20**
COLATN:	Collision with navigation aid	13	2	473,946	
COLBDG:	Collision with bridge	14	2	585,623	
COLCRS:	Collision, crossing	113	10	3,533,480	
COLDOC:	Collision with pier or dock	5	1	78,052	
COLFLO:	Collision with floating object	18	8	1,288,000	
COLFNC:	Collision with fixed object	29	11	1,610,401	1
COLICE:	Collision with ice	7	3	435,100	
COLLDM:	Collision with dike/lock/dam	2	1	109,954	
COLMOD:	Collision, offshore drlng unit	21	3	1,211,974	
COLMTG:	Collision, meeting	68	7	2,080,769	2
COLNEC:	Collision, not elsewhere class.	54	9	3,695,499	4
COLOTK:	Collision, overtaking	61	6	1,653,242	4
COLSPC:	Collision, special circumstance	43	4	732,500	1
COLSUO:	Collision, submerged object	163	53	7,607,446	5
COLUNK:	Collision, unknown	8	1	490,500	
ALLIS:	Allision (coll inv station ves)	120	12	2,810,758	3
DISAPPEARANCE		**20**	**18**	**5,622,300**	**52**
GONENT:	Disappearance, without trace	13	12	4,904,800	36
GONETR:	Disappearance, with trace	7	6	717,500	16

FIRE/EXPLOSION		**599**	**324**	**114,138,504**	**14**
EXPFUF:	Explosion, fuel fire	11	8	911,255	
EXPFUN:	Explosion, no fuel fire	2	1	40,000	
EXPMSF:	Explosion, machinery space fire	32	25	14,133,400	
EXPMSN:	Explosion, mach. space, no fire	5	2	1,011,500	
EXPNEC:	Explosion, not elsewhere class	7	2	336,100	1
EXPPVF:	Explosion, pressure valve, fire	2	2	155,000	
EXPUNK:	Explosion, unknown	5	3	211,000	
FIRELC:	Fire, electrical	38	11	15,269,770	
FIRFUR:	Fire, vessel furnishing	26	7	1,600,470	2
FIRMCS:	Fire, machinery space	321	186	51,094,732	4
FIRNEC:	Fire, not elsewhere classified	84	47	22,695,100	7
FIRSTR:	Fire, vessel structure	61	29	6,551,177	
FIRVFU:	Fire, vessel fuel	5	1	129,000	
FLOODING		**645**	**81**	**30,846,217**	**20**
FOUNDERING		**526**	**362**	**85,432,996**	**96**
GROUNDING		**912**	**155**	**44,354,323**	**13**
GRNDGA:	Grounding, accidental	909	154	43,554,323	13
GRNDGI:	Grounding, inten. w/ dam. haz.	3	1	800,000	

TABLE E-1 *Continued*

Primary Nature of Casualty		Number of Vessel Casualties	Total Vessel Losses	Total Vessel Damages (dollars)	Number of Vessel-Related Fatalities
MATERIAL FAILURE		**2,373**	**100**	**30,101,895**	**6**
MATAGN:	Material failure, aux. gener.	17		50,687	
MATBLG:	Material failure, bilge system	36	4	336,895	
MATCGF:	Material failure, freight hndlg	2	1	0	
MATCGT:	Material failure, tnkr hndlng	1		0	
MATCWS:	Material failure, cooling sys.	165	3	4,643,501	
MATECS:	Material failure, elec cntl sys	14		913	
MATEDS:	Material failure, elec dis sys	67	1	134,139	
MATFCS:	Material failure, feed & cond	1		100	
MATFOS:	Material failure, fuel oil sup	214	4	1,264,405	
MATGTK:	Material failure, ground tackle	19	9	2,970,700	2
MATHCS:	Material failure, hyd ctl sys	19	4	838,137	
MATHDT:	Material failure, hull deter	10	2	296,900	
MATHST:	Material failure, hull, struct	51	15	3,030,590	2
MATLOS:	Material failure, lube oil sup	98	1	1,167,102	
MATLSG:	Material failure, lifesav equip	1		0	1
MATMEN:	Material failure, main engine	642	18	5,320,119	
MATMGN:	Material failure, main gener.	22		66,085	
MATNAV:	Material failure, navig equip	6		56,672	
MATNEC:	Material failure, not else class.	46	6	705,348	
MATPCS:	Material failure, pneum contl	2		130	
MATPRO:	Material failure, propeller	69	2	460,995	
MATRED:	Material failure, reduct gear	267	6	1,067,631	
MATSFT:	Material failure, shaft system	164	5	2,499,950	
MATSWS:	Material failure, salt wat sys	57	3	2,188,378	
MATVNT:	Material failure, vent sys	2		0	
SSFAPS:	Steering sys failure, aux pwr	4		6,750	

SSFCSS:	Steering sys failure, cntl sys	174	6	905,655	
SSFNEC:	Steering system fail, other	37	2	633,106	1
SSFRAS:	Steering sys fail, rdr & shaft	166	8	1,457,007	
OTHER		**606**	**31**	**6,613,355**	**13**
SWAMP:	Swamping	15	5	444,000	3
WAKDMG:	Wake damage	11		27,775	1
WTHRDM:	Weather damage	34	10	2,403,100	1
DISABL:	Disabled	521	14	3,212,330	2
BRGBWY:	Barge breakaway	7	1	4,000	
CARGLD:	Cargo, loss or damage	1		0	
OTHER:	Other	13	1	403,950	2
No primary nature coded		4	0	118,200	4
TOTAL		**6,558**	**1298**	**377,539,947**	**348**

Source: U.S. Coast Guard 1982-1987 CASMAIN data for fishing industry vessels.

TABLE E-2 Documented Fishing Vessel Casualties, 1982-1987, by Primary Cause of Casualty

Primary Cause of Casualty		Number of Vessel Casualties	Total Vessel Losses	Total Vessel Damages (dollars)	Number of Vessel-Casualty-Related fatalities
HUMAN CAUSES		**1,804**	**258**	**99,115,257**	**70**
PBPSASD:	Bypassed avail safety devices	1			
PCALRSK:	Calculated risk	28	8	1,304,700	4
PCRLSNS:	Carelessness	80	9	2,786,079	3
PDEFEQT:	Used defective equipment	9	2	231,900	
PDRUNK:	Intoxication	5		14,700	
PDSGCEX:	Design criteria exceeded	7	4	1,348,100	
PERRJDG:	Error in judgement	203	23	5,179,343	7
PFALACW:	Failed to acct for current	46	7	1,133,480	2
PFALANE:	Failed to use avail nav equip	8	3	358,000	
PFALATR:	Failed to acct for tide/riv sg	14		65,050	
PFALCAP:	Failed to use charts and pubs	9	1	303,475	
PFALEPA:	Failed to est passing agreemen	15	1	279,000	
PFALKPL:	Failed to keep proper lookout	146	24	5,738,200	3
PFALKRC:	Failed to keep to rgt of chan	2		1,450	
PFALPOS:	Failed to ascertain position	110	25	5,954,825	6
PFALRTE:	Failed to use radiotelephone	5	3	603,500	2
PFALRUL:	Failed comply w/rule, reg, pro	34	3	756,652	
PFALSPD:	Failed to proceed at safe speed	14	1	428,500	3
PFALSTP:	Failed to stop	2		14,850	
PFALTY:	Failed to yield right of way	18	1	262,773	
PIMPCCP:	Improper casualty control pro	3	1	220,000	
PIMPCGS:	Improper cargo storage	5	3	331,000	
PIMPFLT:	Improper/faulty lights/shapes	13	2	325,406	1
PIMPLOD:	Improper loading	40	19	16,966,678	28

PIMPMNT:	Improper maintenance	211	13	2,052,212	
PIMPMOT:	Improper mooring/towing	35	10	1,197,330	
PIMPMWS:	Improper/missing whistle/signal	1	1	77,000	
PIMPSCR:	Improper securing/rigging	29	2	888,158	2
PIMPSFP:	Improper safety precautions	37	8	15,027,100	
PINADSP:	Inadequate supervision	3	3	2,552,000	
PINATT:	Inattention to duty	111	12	5,317,458	
PLCKEXP:	Lack of experience	11	3	682,000	
PLCKKNO:	Lack of knowledge	35	1	9,545,100	
PLCKTNG:	Lack of training	3		20,000	
POPERER:	Operator error	424	54	13,915,990	9
POPNFL:	Open flame	3		86,700	
PPVTMNT:	Preventive maintenance not done	10	1	118,400	
PRELFAN:	Relied on floating ATON	1			
PSVCCEX:	Service conditions exceeded	3	1	54,500	
PTIRED:	Fatigue	33	8	2,535,500	
MINSAFT:	Inadequate own/op safety prog	3	1	437,000	
VINFUEL:	Insufficient fuel	34		1,148	
VESSEL CAUSES		**3,042**	**288**	**78,064,944**	**32**
VBRIFA:	Brittle fracture	1		200	
VCGOSHF:	Cargo shift	1	1	125,000	
VCORROS:	Corrosion	15		20,709	
VFATFRA:	Fatigue fracture	17		174,634	
VFLDFST:	Failed fastenings	10	3	60,100	
VFLDMEL:	Failed materials, electrical	291	29	10,909,397	
VFLDMME:	Failed materials, mechanical	1,370	57	20,410,037	2
VFLDMOT:	Failed materials, other	404	36	13,756,118	2
VFLDMST:	Failed materials, structural	324	107	16,393,048	13
VIMPWEL:	Improper welding	8		58,985	
VINADFF:	Inadequate fire-fighting equip	1			
VINADLT:	Inadequate lighting	2		40,000	
VINADMG:	Inadequate/missing guarding	5		10,940	

TABLE E-2 *Continued*

Primary Cause of Casualty	Number of Vessel Casualties	Total Vessel Losses	Total Vessel Damages (dollars)	Number of Vessel-Casualty-Related fatalities
VINADNE: Inadequate navigation equipment	6	1	181,000	3
VINADST: Inadequate stability	18	12	3,395,500	5
VINALUB: Inadequate lubrication	17		75,830	
VINHRSP: Inadequate horsepower	3		41,000	
VNORMLW: Normal wear	56	3	517,496	
VPROFAL: Propulsion failure	110	11	2,341,870	2
VSTRFAL: Steering failure	24	2	356,277	3
VSTRFRA: Stress fracture	3		82,250	
VVIBRAT: Vibration	13		124,741	
VDRGANC: Dragging anchor	34	17	7,186,300	
VFOUPRO: Fouled propellor	291	8	1,341,962	2
MFLTDSG: Faulty design	18	1	461,550	
WEATHER/SEA CAUSES	**232**	**83**	**17,987,853**	**34**
EADVCRT: Adverse current/sea conditions	85	27	7,003,150	14
EADVWTH: Adverse weather	147	56	10,984,703	20
OTHER ENVIRONMENTAL CAUSES	**333**	**80**	**13,676,209**	**6**
ECHNMNT: Channel not maintained	4			
EDEBRIS: Debris	76	10	1,519,342	

EHZBDPC:	Hazardous bridge/dock/ pier com	1		10,000	
EICE:	Ice	8	3	259,750	
ESHOAL:	Shoaling	53	2	789,501	
ESUBOBJ:	Submerged object	161	59	10,505,741	4
ESUCBBV:	Suction bank/bottom/vsl	3		2,055	
EUNCCHZ:	Uncontrollable channel hazard	4	1	150,000	1
EUNMCHZ:	Unmarked channel hazard	16	3	341,420	
MIMADLO:	Improper AID location	1		3,400	
MINADWI:	Inadequate width	1	1	45,000	
MINADWI:	Inadequate AID maintenance	1	1	25,000	1
NVANDAL:	Vandalism	4		25,000	
UNKNOWN CAUSES		**1,147**	**589**	**168,695,684**	**206**
PREVNAT:	Result of previous nature	2		5,500	
NEC:	Not elsewhere classified	131	24	5,634,909	1
UNKNOWN:	Unknown	1,004	564	162,431,019	201
No primary cause coded		10	1	624,256	4
TOTAL		**6,558**	**1298**	**377,539,947**	**348**

Source: U.S. Coast Guard 1982-1987 CASMAIN data for fishing industry vessels.

TABLE E-3 Documented Fishing Vessel Casualties, by Primary Nature of Casualty and Study Region, 1982-1987: All Casualties

Primary Nature of Casualty	Study Region						
	North Atlantic	**South Atlantic**	**Gulf of Mexico**	**West Coast**	**Alaska**	**Other**	**Total**
Capsizing	13	6	26	43	34	16	138
Collision	158	45	284	104	85	63	739
Disappearance	2	0	3	11	3	1	20
Fire/Explosion	123	55	117	107	113	84	599
Flooding	176	50	150	138	60	71	645
Foundering	112	32	106	123	84	69	526
Grounding	161	162	150	165	195	79	912
Material Failure	733	91	192	928	160	269	2373
Other	210	35	77	187	36	61	606
Total	**1,688**	**476**	**1,105**	**1,806**	**770**	**713**	**6,558**

Source: U.S. Coast Guard 1982-1987 CASMAIN data for fishing industry vessels.

TABLE E-4 Documented Fishing Vessel Casualties, by Primary Nature of Casualty and Study Region, 1982-1987: Total Vessel Losses

Primary Nature of Casualty	Study Region						Total
	North Atlantic	South Atlantic	Gulf of Mexico	West Coast	Alaska	Other	
Capsizing	5	5	13	33	30	8	94
Collision	20	11	50	25	9	18	133
Disappearance	2	0	3	10	3	0	18
Fire/Explosion	49	38	63	61	63	50	324
Flooding	16	5	16	17	21	6	81
Foundering	65	22	68	95	70	42	362
Grounding	16	20	23	33	49	14	155
Material Failure	13	9	24	22	15	17	100
Other	4	4	9	4	8	2	31
Total	**190**	**114**	**269**	**300**	**268**	**157**	**1,298**

Source: U.S. Coast Guard 1982-1987 CASMAIN data for fishing industry vessels.

TABLE E-5 Documented Fishing Vessel Casualties, by Vessel Length and Study Region: All Casualties

Vessel Length (feet)	Study Region						
	North Atlantic	South Atlantic	Gulf of Mexico	West Coast	Alaska	Other	Total
26-49	586	177	399	1173	416	299	3050
50-64	349	89	254	302	99	124	1217
65-78	433	165	388	201	78	195	1460
79 or more	316	42	61	95	173	92	779
Unknown or <25	4	3	3	35	4	3	52
Total	**1,688**	**476**	**1,105**	**1,806**	**770**	**713**	**6,558**

Source: U.S. Coast Guard 1982-1987 CASMAIN data for fishing industry vessels.

TABLE E-6 Documented Fishing Vessel Casualties, by Vessel Length and Study Region: All Casualties—Total Vessel Losses

Vessel Length (feet)	Study Region						
	North Atlantic	**South Atlantic**	**Gulf of Mexico**	**West Coast**	**Alaska**	**Other**	**Total**
26-49	62	42	102	192	157	70	625
50-64	48	30	69	49	24	34	254
65-78	51	37	88	34	27	38	275
79 or more	29	3	9	22	58	14	135
Unknown or <25	2	1	3	2	1	9	
Total	**190**	**114**	**269**	**300**	**268**	**157**	**1,298**

Source: U.S. Coast Guard 1982-1987 CASMAIN data for fishing industry vessels

TABLE E-7 Primary Causes of Non-Vessel-Casualty-Related Commercial Fishing Fatalities, 1982-1987

HUMAN CAUSES	**82**
Carelessness	15
Failure to use PFD	1
Improper loading/storage	0
Improper supervision	1
Inadequate training	3
Misuse of equipment	1
Psychological factors	3
Unsafe movement (self)	10
Unsafe practice (another)	3
Unsafe practice (self)	34
Intoxication (self)	10
Narcotics	1
VESSEL CAUSES	**20**
Slippery Deck	2
Equipment failure	7
Chemical reaction	1
Improper maintenance	1
Insufficient ventilation	2
Material failure	2
Physical factors	5
ENVIRONMENTAL CAUSES	**10**
Weather	10
UNKNOWN	**98**
Unknown	79
Missing	3
Definition unknown	1
Not elsewhere classified	11
Vessel casualty and adjustment*	4

APPENDIX

F

Coast Guard Compliance Examinations

Coast Guard compliance examinations constitute the principal method by which uninspected fishing vessels are exposed to federal checks for mandated equipment and adherence to federal laws and regulations. This appendix assesses their role in improving safety.

THE NATURE OF COMPLIANCE ACTIVITY

Compliance examinations are most closely associated with enforcement of laws and regulations pertaining to recreational boating safety, fisheries management, conservation, marine pollution, and customs. The Coast Guard refers to this activity as maritime law enforcement (MLE), enforcement of laws and treaties (ELT), and sometimes inspections, although four distinct Coast Guard programs are involved. But in practice, the term "boardings" is most frequently used, reflecting the process by which "compliance examinations" are conducted. In this appendix, boardings refers to the process, and compliance examinations to the function. This terminology is also applied to distinguish this activity from the Coast Guard's vessel inspection program.

Boardings are conducted by the Coast Guard to detect and suppress violations of all federal laws. The nature and frequency of compliance examinations of uninspected fishing vessels during boardings varies depending upon Coast Guard operational priorities, resource availability, and funding. They may be random, the result of an apparent violation or problem observed by the Coast Guard, conducted to satisfy the objectives of several Coast Guard programs, or coincidental with other operational activities. They may be conducted under way or dockside. The Coast Guard attempts to minimize inconvenience to vessels boarded. However, the Coast Guard and the National Marine Fisheries Service (NMFS) consider boarding of vessels on the fishing grounds while engaged in fishing essential to enforcement (see Sutinen et al., 1989a,b).

Records of prior compliance examinations are generally not accepted by Coast Guard boarding officers as current evidence of compliance. A full examination is usually conducted to ensure that the vessel's compliance status has not changed. NMFS officials sometimes accompany Coast Guard personnel during examinations conducted for fisheries management and conservation (see Sutinen et al., 1989a,b; Chandler, 1988; National Oceanic and Atmospheric Administration [NOAA], 1986).

Inspected commercial vessels and uninspected commercial vessels that require a licensed operator (e.g., uninspected passenger and towing industry vessels) are not normally subjected to comprehensive enforcement activities while under way. Preinspections may be conducted while a vessel is under way, but more thorough compliance examinations are usually conducted while the vessel is moored at an industrial facility, often to determine compliance with marine pollution regulations as in the case of tankers or tank barges. Underway boarding of these vessels for general law and regulatory enforcement is an exception, usually where there is evidence or suspicion of a violation or illegal activity like drug smuggling. Thus, while all commercial vessels may be boarded under way, only uninspected fishing vessels are routinely stopped while actively engaged in their trade.

If the Coast Guard establishes a presail inspection requirement for fishing vessels, under present compliance examination policies fishing vessels could be subject to both mandatory inspections prior to operation and continued random compliance examinations. Some underway boardings of fishing vessels are expected to remain standard Coast Guard procedure regardless of the outcome of rulemaking required by the Commercial Fishing Industry Vessel Safety Act of 1988 (CFIVSA), because these enforcement measures are needed to oversee compliance with fisheries management, conservation, and smuggling laws. Whether a valid inspection record of some form will be considered evidence of compliance with safety equipment regulations by boarding officers conducting operational examinations is an unresolved Coast Guard policy issue.

OBJECTIVES OF COMPLIANCE EXAMINATIONS

Ideally, the focus of safety initiatives would be motivating changes that could overcome or mitigate safety problems by leading to safer vessels, operations, and work practices rather than compliance with laws and regulations. Compliance examinations could serve as a check on safety performance, with passing them an incidental and natural by-product of effective safety programs.

Conceptually, this could help ensure that compliance programs do not become the end rather than the means.

The objectives of Coast Guard maritime law enforcement boardings of recreational and uninspected vessels are to deter unsafe operation, to detect vessel violations, and to educate the maritime public. Boarding officers are charged to promote safety by pointing out and explaining potentially dangerous conditions, whether or not they are contrary to laws and regulations (U.S. Coast Guard [USCG], 1986a). The effectiveness of compliance examinations in achieving fisheries management objectives and improving safety aboard uninspected fishing vessels has not been evaluated by the Coast Guard.

In practice, the general focus of boarding activity has been on compliance rather than education. This is attributed to the wide range of boating safety, uninspected vessel, fisheries, conservation, and drug laws that are enforced as well as the operating conditions. The compliance focus to some degree appears to reflect the experience of relatively young technical personnel who conduct many of the boardings and who frequently have significantly less sea time (although not necessarily less knowledge) than the mariners with whom they interact. The myriad laws and regulations that boarding officers must address lead to using checklists to ensure that the range of regulations they are enforcing are adequately addressed.

The boarding process can be as short as 20-30 minutes to more than 12 hours if a haul back is required to inspect the fishing gear and catch, as may be the case for factory trawler operations. A full drug enforcement inspection can require movement of fishing gear, which is frequently not an easy task, especially if there is a heavy deck load. All this is not conducive to on-the-spot education of the recipients. Thus, the boarding encounter aboard fishing vessels tends to be adversarial in nature. However, it may still be possible for skilled boarding officers, in situations where the adversarial nature of the event is less pronounced, to direct attention to actual or potential safety problems and encourage corrective action.

THE EXTENT AND RESULTS OF COMPLIANCE EXAMINATIONS

During fiscal year 1989 (FY 89), the Coast Guard's Summary Enforcement Event Report (SEER) data base recorded 8,176 of 35,622 total boardings for maritime law enforcement purposes involving uninspected fishing vessels. Over 2,200 of these were repeat boardings, a tactic used for fisheries enforcement purposes (see Sutinen et al., 1989a,b). SEER data do not distinguish between boardings while the vessel is fishing, in transit, or moored or anchored. In practice, most of these boardings are performed while the vessel is under way and often while fishing, especially when it is essential for determining whether correct net sizes are in use or the proper size and types of species are being taken.

Thirty-eight percent of fishing vessel boardings resulted in "boating safety" violations. About 41 percent of all boardings recorded in SEER during FY 89 resulted in a boating safety violation. While the exact nature of these violations could not be determined, the data suggest that there is no significant difference in compliance with basic federal regulations between recreational boaters and commercial fishermen. Compliance with basic safety regulations that predate the CFIVSA is thought to be reasonably high. Coast Guard personnel familiar with the operational examinations of uninspected fishing vessels suggest that no more than 10-15 percent of violations were related to deficiencies in basic safety and survival equipment. About 11 percent of violations reported during a special operational examination program targeting the Sitka sac roe herring fishery in FY 89 were related to safety equipment deficiencies. A case study of this program is presented later in this appendix. Similar safety data were not available for other regions.

Of 144 vessels seized for drug violations during FY 89, 61 were uninspected fishing vessels. SEER recorded 95 arrests associated with these seizures. It should be noted that a "zero tolerance" policy was in effect during the period. FY 89 seizures were up slightly from FY 87 and 88, while arrests were somewhat lower. Most of the seizures and arrests were associated with marijuana. Data on abuse of alcohol by fishermen were not available. SEER data suggest that the majority of fishing vessels were not involved in transporting contraband. However, substance abuse (both alcohol and drugs) was mentioned frequently during the regional assessments commissioned for this study. No data were available to indicate the degree to which substance abuse in the fishing industry compares with trends in other occupations or society at large. The general perception is that use of illegal drugs is more closely associated with highly transient crews than with vessels exhibiting a high degree of crew stability.

During FY 89, using an estimated national fleet size of 111,000 vessels, an estimated 5-6 percent of uninspected fishing vessels nationally were subjected to Coast Guard equipment checks. About 2 percent were subjected to multiple examinations. Up to four boardings of the same vessel in 1 year are known to have occurred. Because of the enhanced boarding activity in selected Alaskan waters, a slightly higher percentage of uninspected fishing vessels there were exposed to direct Coast Guard checks of required equipment than elsewhere. Overall, boarding information is not conclusive, but suggests that uninspected fishing vessels generally comply with safety and survival equipment requirements in effect prior to June 1990.

COURTESY MARINE EXAMINATIONS

The Coast Guard does not offer courtesy examinations through organized Coast Guard units. However, uninspected fishing vessels are eligible for courtesy marine examinations (CMEs) offered by the Coast Guard Auxiliary, a volunteer organization of private citizens sponsored by the Coast Guard. Auxiliary activity is almost exclusively oriented toward the recreational public. Uninspected fishing vessel use of the CME service is considered negligible by the Coast Guard. The number of commercial fishermen who may have attended Coast Guard Auxiliary, U.S. Power Squadron, or other such public education courses is not known.

CASE STUDY: SITKA SAC ROE HERRING FISHERY COMPLIANCE EXAMINATION PROGRAM

The fishing industry is a major element of the Alaska state economy, and fishing there is front-page news. Well-publicized vessel losses and multiple fatalities in Alaskan waters have brought continuing attention to safety and survival at sea. Since 1988, the Coast Guard has conducted an intensive operational examination program just prior to and during the spring Sitka sac roe herring season in order to promote safety. During 1989, Coastguardsmen conducting the examinations were specially trained for the operation. A combination of underway and dockside boardings was employed. Underway boarding locations were periodically adjusted to offset changes in fishing vessel operating patterns intended to minimize exposure to boardings. As a result, the entire sac roe herring fleet operating in southeastern Alaskan waters was exposed to a comprehensive examination of vessel compliance with environmental and safety regulations. A checklist of 48 items was used. However, safety issues were stressed during boarding. Incidental boardings of uninspected fishing vessels from other fisheries were also recorded (USCG, 1988b, 1989b).

The sac roe herring fishery was chosen for this special local program because it is the first fishery to open each year in the Alaskan panhandle and brings a large number of vessels into the principal southeast fishing ports. Between 150 and 250 boats were boarded annually; 244 vessels were boarded during the 1989 fishery. Over 61 percent of boardings detected one or more violations; however, a high incidence of compliance with Coast Guard requirements for lifesaving equipment was observed. Of 294 violations, only 20 were for personal flotation devices (PFDs) and 13 for fire extinguishers; most of the rest were associated with marine sanitation devices, incorrect installation and use of oily water separators (where required), and other environmental regulations (USCG, 1989b). A monitoring program for actual safety problems experienced during the fishery has not been established, so it is not apparent whether the program has had residual effects in reducing the incidence of casualties or fatalities.

The number of violations decreased annually during the program's first 2 years as fishermen learned of it and prepared their vessels accordingly. The Coast Guard observed that over 90 percent of discrepancies recorded during the 1988 season had been corrected by the opening of the 1989 fishery. Additionally, every vessel boarded in Sitka had sufficient immersion suits on board and at least one emergency position-indicating radio beacon. A similar program was conducted for Prince William Sound during spring 1988. A dockside boarding program was conducted over a 10-day period in Dutch Harbor, Alaska, also during 1988. The Coast Guard indicates that these programs will continue into the near future under special funding for Alaska-based fishing vessel safety projects. Similar programs for other regions are not funded.

Because the Coast Guard has established the credibility of this local boarding program, it appears uniquely positioned to shift its emphasis to dockside boardings at central locations in advance of the fishery opening, should the Coast Guard choose to do so. This type of approach could lead to detection and correction of safety deficiencies and some savings in Coast Guard operating costs. Dockside safety seminars could also be attempted, such as practical instruction in donning immersion suits. If dockside safety inspection records were accepted as evidence of compliance with safety regulations in lieu of on-site verification while in transit or fishing, underway boardings could be reduced to levels necessary to ensure adherence to fisheries regulations and for correction of obvious safety deficiencies. This could reduce boarding times, minimizing inconvenience to the fishermen.

SUMMARY

Boardings do not appear to incorporate the kind of mutual cooperation and accountability for safety that might lead owners, operators, or crew to address safety as a continuing need that is in their self-interest rather than as an item-by-item checkoff on a list of federal requirements. Compliance examinations do not assess attitudes; fishing, navigational, safety, or survival skills; or, in all but the more obvious circumstances, the material condition of the vessel and machinery. At best, each boarding represents a "snapshot" of the vessel's compliance status with applicable federal laws and regulations at the time of boarding.

Compliance by uninspected fishing vessels with Coast Guard regulations in place prior to June 1990 for onboard quantities of Coast Guard-approved safety and survival equipment appears high nationwide. However, boarding results do not necessarily reflect how well a vessel will comply with laws and regulations and with accessibility, maintenance, and use criteria for required equipment while not under direct Coast Guard observation, which is most of the time. As structured, underway boardings and compliance examinations do not appear to effectively improve safety attitudes and procedures that underlie the overwhelming number of vessel casualties, injuries, and fatalities.

The Coast Guard has established a highly credible compliance program for the Sitka sac roe herring fishery, principally through extensive underway boardings. The heavy commitment of Coast Guard resources has resulted in a steady decline in observed deficiencies in required safety equipment. An associated improvement in the overall safety status of examined vessels is a key objective. However, the actual effectiveness of the boarding program in promoting safety practices that prevent accidents or mitigate their effects has not been determined through performance monitoring.

APPENDIX

G

Inspection of Fishing Industry Vessels

The Coast Guard administers several programs intended to promote marine safety on federal waters, among them commercial vessel safety. Major program elements include vessel documentation, vessel inspection, licensing of personnel, and pilotage. Currently, the agency formally inspects 12 different types of merchant vessels, including tankers, cargo ships, and passenger vessels of 100 net tons or greater. Of fishing industry vessels (i.e., fishing vessels, fish tender vessels, and fish processing vessels), only one—a large processor—is known to be inspected.

A fundamental question is whether safety can be improved through vessel inspection. This was answered affirmatively by the analysis presented in Chapters 3 and 4. Major findings include the following:

- About 85 percent of casualties to documented fishing industry vessels from known causes are closely associated with failed material and equipment.
- Lack of design and stability problems contribute to a significant number of vessel total loss incidents.
- Over 65 percent of all Coast Guard search and rescue (SAR) cases involving fishing industry vessels have been attributed to failure of hulls, propulsion equipment, or other machinery.

This appendix examines technical characteristics of vessel inspection requirements imposed by 46 U.S.C.A. Chapter 33 and the efficacy of extending it to some or all uninspected fishing industry vessels to improve vessel and equipment fitness for service.

STATUTORY AND REGULATORY REQUIREMENTS

Vessel Registration

The national fishing industry fleet consists of federally documented vessels authorized to participate in the fishing trade and vessels bearing state numbers used for commercial fishing. Vessel documentation and state numbering are registration programs. Registration records do not include data on how many vessels actively fish or statistics about their operations, indications about their physical (i.e., material) condition, hull forms, or fishing gear.

Documentation

A Certificate of Documentation (not to be confused with a Certificate of Inspection) is required to operate vessels of at least 5 net tons in certain trades and must be renewed annually. It is evidence of vessel nationality and, with exceptions, permits vessels to be subject to preferred mortgages. An official number is assigned to the vessel, which remains with the vessel while it is under U.S. registry regardless of name or ownership changes. The number must be permanently marked inside the vessel's structure.

The Certificate of Documentation may be endorsed for certain trades. The endorsement is commonly referred to as a "license." A vessel's document may be endorsed for many eligible trades (e.g., registry, coastwise, Great Lakes, fisheries, or recreational), but it must have a fisheries license endorsement if it is going to engage in fishing. This endorsement entitles the vessel to fish and land its catch, wherever caught, in the United States. It pertains to the legal privilege to engage in the fishery trade, is issued without regard to a vessel's suitability or fitness for such service, and does not replace vessel licenses required by other federal or by state authorities to harvest specific species. By federal law, a vessel permanently loses its fishery license endorsement if it undergoes rebuilding outside the United States (46 U.S.C.A. Chapter 12).

The following vessels, if at least 5 net tons and wholly owned by a U.S. citizen, are eligible for a fishery license endorsement:

- vessels built in the United States,
- wrecked vessels,
- forfeited vessels,
- captured vessels, and
- vessels granted fisheries privileges by special legislation.

In March 1990, documentation data recorded in the Coast Guard's Marine Safety Information System (MSIS) indicated there were about 30,000 documented vessels with fishery endorsements. However, a vessel's actual use in the fishing industry is not monitored by Coast Guard data bases or MSIS. Thus, the

PRINCIPAL FISH TENDER AND PROCESSING VESSEL OPERATIONS

A large number of fishing industry vessels are employed as fish tender vessels in Alaska, notably in the salmon fisheries. Fishing vessels often make ideal platforms for this type of activity. For example, crab vessels with chilled seawater tanks make excellent platforms for transporting fresh salmon from remote locations to shoreside cold storage and processing plants. Some fishing industry vessels are configured for both catching and processing fish. Some fish processors are also used periodically as fish tenders.

In recent years, about 1,400 vessels have been issued permits by the Commercial Fisheries Entry Commission in Juneau, Alaska, for tender/packer operations. During 1987, about 37 percent were 40 feet or less in length and about 43 percent were between 41 and 78 feet. The remaining 20 percent were 79 feet or longer and potentially subject to load line regulations if they are not fishing vessels chartered as "part-time" tenders (i.e., a fishing vessel is one that fishes more of the year than it tenders).

National Marine Fisheries Service and Commercial Fisheries Entry Commission data indicate that there are about 200 vessels that process fish. The Seventeenth Coast Guard District, Juneau, Alaska, estimates that about 50 of these appear to be configured solely as floating processors. The others may have catching or tender capabilities. It is not known how many fishing industry vessels cross over between the three categories or how many are subject to load line regulations.

number of vessels principally configured as fish tendering or processing vessels cannot be directly determined from Coast Guard data (see box).

State Numbers

The remainder of the fishing industry fleet, about 80,000 vessels, bear state numbers issued by state authorities (except in Alaska, where state numbers are administered by the Coast Guard). State numbers are required for virtually all boats by federal law, and these data provide a way to estimate the general population of vessels below 5 net tons. However, there is no fisheries endorsement akin to that required for use of documented vessels in the fisheries trade.

Tonnage Measurement

Only vessels measuring 5 net tons and larger are eligible for documentation; therefore, evidence of tonnage is required. Tonnages may be physically determined (Convention, Standard, or Dual Tonnage) or calculated based on an owner's statement of dimensions (Simplified Measurement). With certain exceptions, this latter method is available as an owner's option for all vessels less than 79 feet in length (46 CFR 69, Subpart E).

As of January 1, 1986, new vessels at least 79 feet overall in length that will be documented or that engage on a foreign voyage are required to be measured under the Convention Measurement Method (46 CFR, Subpart B). This method is based on the rules of the 1969 Tonnage Convention.

Tonnages are used extensively to regulate vessels. (The Coast Guard employs tonnage in its vessel inspection regulations and as a basis for personnel licensing and for vessel manning [see Wiese, 1978].) A vessel required to be measured under the Convention Measurement Method may, at the option of the owner, be measured also under the preexisting national systems. Tonnages derived under those systems (Standard and Dual Tonnage Methods [46 CFR 69, Subparts C and D]) may be used when a vessel is regulated domestically on its tonnage. In some cases, the regulatory tonnage may be used also for limited regulatory applications under the Safety of Life at Sea (SOLAS) Convention, International Convention for the Prevention of Pollution (MARPOL), and Standards for Training, Certification, and Watchstanding (STCW) Convention.

American Bureau of Shipping (ABS) was delegated the ministerial functions of physically measuring commercial and recreational vessels of any size and issuing tonnage certificates. ABS forwards tonnage certificates to a Coast Guard office where the vessel documentation records are maintained. For vessels less than 79 feet long, the simplified method is incorporated in the vessel documentation process.

Vessel Inspection

While all commercial vessels of at least 5 net tons are required to be documented, "vessel inspection" is not required for all documented vessels. Vessel inspection as used in this report refers to visits aboard documented vessels by Coast Guard marine safety personnel to verify compliance with regulations issued under the authority of 46 U.S.C.A. Chapter 33.

Background

Merchant vessel inspection is well established in law and regulation and includes a large number of small coastal and harbor vessels. Title 46 of the U.S. Code comprises the U.S. shipping laws; Title 46 of the U.S. Code of

VESSEL INSPECTION POLICY

Vessel Inspections Vessel inspection policy is developed to ensure the uniform and consistent administration and enforcement of federal laws and regulations designed to protect individuals, their private property, and the marine environment from the consequences of incidents involving materially unsafe vessels. Inspection is the process of examining a vessel to determine its reasonable, probable compliance with minimum safety standards over a projected period of time. A Certificate of Inspection (COI) attests to that reasonable probability.

Other Examinations These are examinations that may be made while the vessel is in operation and include all other inspections and visits to the vessel by Coast Guard marine safety personnel. Failure to meet certain standards may result in withdrawal of, or issuance of amendments or requirements to, the COI issued to a vessel; action taken against its operating personnel under suspension and revocation proceedings; or civil penalties imposed against the vessel's owner or operator or other responsible party. (Note: "Other examinations" of inspected vessels generally do not include maritime law enforcement boardings [compliance examinations], although such boardings are within the Coast Guard's jurisdiction and authority.)

Source: USCG Marine Safety Manual, Volume II (COMDTINST 16000.7).

Federal Regulations delineates applicable vessel inspection regulations. The Coast Guard's *Marine Safety Manual, Volume II* (U.S. Coast Guard [USCG], 1985), serves as the principal document guiding administration of material inspection regulations. Refinements to inspection laws, regulations, and policy have been heavily influenced by lessons learned from past marine disasters.

General Inspection Requirements

U.S. shipping laws and regulations impose vessel inspection requirements based on tonnages and other parameters and periodic in-service checks and reinspections. Vessel inspection policy has two components—"inspections" and "other examinations" (see box). Vessel inspections can be extensive, and there are significant resource implications to both the government and fishing industry that could make implementation of such a program very expensive and difficult. This is suggested by the nature and scope of vessel inspection discussed later.

Administration of vessel inspection requirements constitutes an action-forcing mechanism for motivating and achieving compliance. Satisfactory completion leads to a Coast Guard Certificate of Inspection (COI), which attests that the vessel has satisfied applicable criteria. A successful vessel inspection does not necessarily mean that a vessel is safe, however. It does mean that masters or individuals in charge are provided with a vessel that has been certified to minimum standards deemed essential for safe operations. With periodic preventive and corrective maintenance, this fitness for service is expected to endure for the period for which the COI is valid.

Fishing Industry Vessels Subject to Inspection

Over 99 percent of fishing industry vessels have not been regulated for safety except for general safety equipment and navigational safety requirements applicable to all uninspected vessels. Only fish processing vessels of more than 5,000 gross tons and fish tender vessels of more than 500 gross tons are subject to formal vessel inspection (46 U.S.C.A. §3301). Only one fish processing vessel, a converted container ship, has been identified by the Coast Guard as subject to vessel inspection. Virtually all other fishing industry vessels fall below thresholds that would trigger the vessel inspection requirement.

Fish tender vessels and fish processing vessels 79 feet or longer are subject to load line regulations and associated inspection requirements (except where grandfathered) to ensure the integrity of the hull and all watertight and weathertight closures (46 U.S.C.A. Chapter 51). Fish processing and tendering vessels, with few exceptions, conduct their operations in the Alaska region (see box, p. 224). Fishing vessels are exempt from load line regulations (46 U.S.C.A. §3302).

Commercial Fishing Industry Vessel Safety Act (CFIVSA) of 1988

The CFIVSA (P.L. 100-424) provides the Coast Guard with authority to bring the commercial fishing fleet under more rigorous safety regulation. The CFIVSA requires this study to recommend whether or to what degree a vessel inspection program should be used to improve the safety of fishing industry vessels. The act applies in varying degrees to all uninspected fishing, fish processing, and fish tender vessels. This includes federally documented and state-numbered vessels not subject to vessel inspection. As a practical matter, the distinction between the three categories of vessels covered by the CFIVSA is sometimes tenuous as a result of multiple uses.

Safety requirements for all fishing industry vessels operating beyond the Boundary Line (essentially those that operate on coastal waters and high seas) or with more than 16 individuals aboard were increased by the CFIVSA. The act established authority for even more rigorous safety requirements for fishing industry vessels that operate beyond the Boundary Line with more than 16 individuals aboard that were built or converted after December 31, 1989. The act also mandated certification to classification society survey and classification requirements for uninspected fish processing vessels built or converted after July 27, 1990. Furthermore, the act requires examination of fish processing vessels once every 2 years. It also requires the Coast Guard to conduct a separate study of fish processing vessels that are not classed or surveyed by an organization approved by the Secretary of Transportation. The purpose of the fish processing vessel study is to determine what hull and machinery requirements should apply to ensure adequate maintenance and safe operation at sea (P.L. 100-424).

Occupational Safety and Health

To the extent that the Coast Guard has no regulations dealing with particular occupational hazards on uninspected fishing industry vessels, the Occupational Safety and Health Administration (OSHA) regulates occupational safety and health, as set forth in section 4(b)(1) of the Occupational Safety and Health Act of 1970. When OSHA conducts inspections aboard fishing industry vessels, the Coast Guard is advised in advance and reserves the right to accompany OSHA compliance officers. Specific occupational hazards noted by OSHA are referred to the Coast Guard if the hazards are addressed by Coast Guard regulations. Otherwise, the hazards are cited under OSHA regulations.

The Coast Guard and OSHA have maintained liaison concerning promulgation of safety regulations in response to the CFIVSA. However, there are areas for which the Coast Guard has not proposed regulations for occupational safety and health. In particular, the occupational safety and health of industrial personnel working in industrial facilities that process fish aboard fishing industry vessels is anticipated to remain an OSHA responsibility under 29 CFR 1910, General Industry Standards. Issues addressed for afloat industrial facilities include: machinery and machine guarding, ergonomics, noise, ventilation, personnel protective equipment, materials handling and storage, electrical systems, and lighting conditions.

OSHA conducts "programmed" (i.e., scheduled) safety inspections of fishing industry vessels that process fish products and have aboard 11 or more employees. Pursuant to appropriations limitations, OSHA is precluded from conducting programmed safety inspections of worksites in the fishing industry with 10 or fewer employees. Regardless of the number of worksite employees, OSHA has authority to conduct health inspections aboard uninspected fishing industry vessels in response to imminent danger situations, fatalities, catastrophes, complaints, referrals, and follow-ups of a previous inspection. A nominal number of unprogrammed inspections are done annually. Specific sources of notification and information leading to an unprogrammed inspection include:

accidents reports, media reports, employee complaints, and referrals from other agencies, companies, or inspectors.

A major limitation on OSHA inspections is that OSHA has jurisdiction over vessels only when they are operating within the limits of state territorial seas. Generally, this is within 3 nautical miles of shore, except for the Gulf Coast of Florida and Texas, where 3 marine leagues, or about 9 nautical miles, defines OSHA's seaward jurisdictional limit. Additionally, vessels, including some factory trawlers or other catcher/processors, that simply head and gut fish prior to onboard storage are not defined as "fish processors" by 46 U.S.C.A. §2101, although they are generally treated as such by some states. OSHA does not systematically regulate the safety of industrial facilities aboard these vessels, although unprogrammed inspections may be conducted as previously noted.

ADMINISTRATION OF VESSEL INSPECTION

Coast Guard objectives for administration of vessel inspection are shown in the box (USCG, 1985). The owner is responsible for the safety of the inspected vessel. The master or individual in charge, often referred to as the operator, is responsible for operating it safely (the term operator is also commonly used to refer to a company that charters a vessel). Vessel inspection does not remove the owner's responsibility for vessel seaworthiness; it verifies fitness for service. The Coast Guard uses vessel inspection as a strategy to check the diligence of vessel owners in ensuring that the vessel is properly constructed, configured, equipped, and maintained. The owner must also keep the Coast Guard apprised of major alterations or repairs. Vessel owners are responsible for ensuring that their management, employees, and crews adhere to regulated safety requirements. However, owners sometimes fail to meet their obligations, necessitating enforcement measures.

Most fishing industry vessels fall into the harvest sector, which is where most vessel and personnel casualties are recorded. Therefore, the remainder of this appendix is directed toward vessels configured principally as fishing vessels. However, the basic analysis is relevant for vessels that cross over into fish tender operations or are configured as catcher/processors, including head-and-gut boats. The basic analysis is also relevant to the construction, maintenance, and outfitting of fish tender and processing vessels. It does not apply to their industrial processing capabilities.

The Vessel Inspection Process

Coast Guard procedures include approval of plans and on-site inspection of items during and after construction (or conversion), installation of equipment, or repairs. The on-site inspection verifies that the new installation or repairs follow approved plans and meet federal regulations.

COAST GUARD VESSEL INSPECTION OBJECTIVES

The Coast Guard administers vessel inspection laws and regulations to promote safe, well-equipped vessels that are suitable for their intended service without placing an unnecessary burden upon the economic and operational needs of the marine industry. In determining inspection requirements and procedures, inspection personnel must recognize and consider the following factors:

- The burden for proposing acceptable repairs rests upon the vessel's owner, not upon the repair facility or inspector.
- Delays to vessels are costly, and the need for a delay must be balanced against the risks imposed by continued operation of the vessel.
- Certain types of construction, equipment, and/or repairs are more economically advantageous to the vessel operator and can provide the same measure of safety.
- Some repairs can be safely delayed and can be more economically accomplished at a place and time proposed by the vessel operators.
- The overall safety of a vessel and its operating conditions, such as route, hours of operations, and type of operation, should be considered in determining inspection requirements.
- Vessels are sometimes subject to operational requirements of organizations and agencies other than the Coast Guard.
- A balance must be maintained between the requirements of safety and needs for practical operations. Arbitrary decisions or actions that contribute little to the vessel's safety and tend to discourage the construction or operation of vessels must be avoided.

Source: USCG Marine Safety Manual, Volume II (COMDTINST 16000.7).

Vessel inspection addresses the following:

- design;
- construction;
- equipment;
- stability;
- manning requirements;
- maintenance;
- route; and
- service (i.e., the allowable trades in which the vessel may engage).

Plan Review

Designs of vessels subject to inspection must be approved by the Coast Guard before the owner or builder can legally begin construction or apply for inspection. Depending on a vessel's size, its plans are approved either by the Coast Guard Marine Safety Center in Washington, D.C., or the cognizant Officer in Charge of Marine Inspection at the port nearest where the vessel is being built (46 CFR 91.20).

In the case of Mobile Offshore Drilling Units (MODUs), the Coast Guard accepts designs of certain "industrial systems" certified by a registered professional engineer as allowed under 46 CFR 58.60-11. If a vessel is to be "classed," many plans may be reviewed by the ABS under an agreement with the Coast Guard enabled by 46 U.S.C.A. §3316 (USCG, 1982). For all other commercial vessels subject to inspection, designs are reviewed by Coast Guard technical personnel at the Marine Safety Center or by marine inspectors under the cognizant Officer in Charge, Marine Inspection.

The following features typically require review to verify that standards and requirements have been met (46 CFR 91.55):

- specifications;
- general arrangement plan;
- hull structure;
- subdivision and stability;
- fire control;
- marine engineering;
- electrical engineering;
- lifesaving equipment; and
- crew's accommodations.

Inspection

After plans are approved, an "Application for Inspection of U.S. Vessel," Form CG-3752, is filed with the cognizant Officer in Charge, Marine Inspection. The form serves as formal notification that the owner, master, or builder desires the vessel to be inspected, when it will be ready, and its location. A Coast Guard marine inspector, an ABS marine surveyor, or both will attend the vessel throughout its construction to see that it is constructed in accordance with approved plans, Coast Guard regulations, and ABS rules. When construction is completed, tests and inspections are conducted to ensure that systems such as steering, propulsion control, and fire control perform to prescribed standards (46 CFR 91.25).

During the final days of outfitting, an inclining experiment is conducted to determine the vessel's intact stability. The experiment consists of placing specified weights at prescribed locations and measuring to verify theoretical

TRIM AND STABILITY BOOK INFORMATION

- Instructions for using the book
- Table of contents and index
- Vessel general description
- Vessel schematic showing watertight bulkheads, closures, downflooding angles, and similar key information
- Hydrostatic curves
- Tank sounding tables
- Loading restrictions
- Examples of loading criteria
- Rapid and simple means for evaluating other loading conditions
- Description of stability calculations
- Precautions to prevent unintentional flooding
- Information regarding use of crossflooding fittings
- Amount and location of fixed ballast

Source: USCG Marine Safety Manual, Volume II.

stability data. The validated or corrected data (see box) are then incorporated into a Trim and Stability Book that is reviewed by the Coast Guard and must remain onboard the vessel. It provides information needed by the master or operator to readily ascertain the vessel's stability under varying loading conditions. All manned vessels reviewed for stability receive a U.S. Coast Guard stability letter which, in the case of certain small vessels, such as small passenger vessels, may contain sufficient guidance to obviate the need for a Trim and Stability Book.

Certificate of Inspection (COI)

The COI is issued only after all applicable regulations have been complied with. This includes ensuring that required lifesaving equipment is on board and in good condition, that the fire and bilge pumping systems work, and that electrical systems are of the correct size and material. Vessels with a COI are continuously monitored using MSIS, until the owner surrenders the COI. When this occurs, the Coast Guard drops the vessel from the active list in MSIS.

The Coast Guard issues a COI once all of the above steps have been successfully completed. It certifies that the vessel is fit for the route and service for which it is intended, provided that it is operated according to the COI's terms and conditions. A COI, subject to annual inspection, is valid for 1 to 3 years, depending on the vessel, and must be renewed prior to expiration.

A typical COI specifies amounts and types of safety equipment that must be carried, cites dates various examinations—such as drydocking—were conducted and when the stability book or letter were issued, manning requirements, numbers of persons allowed on board, and the route and trades in which the vessel may engage. The COI is the final document issued to a new vessel when construction and inspection are completed. It must be kept current in order for insurance, charter agreements, and similar contractual arrangements to remain in force. The COI is usually posted in the pilothouse (46 U.S.C.A. §3309-3314).

Recertification and Reinspection

Inspected vessels are required to be recertified and reinspected periodically (46 U.S.C.A. §3307-3308). For example, 65-foot small passenger vessels, comparable in size to many fishing vessels, are required to be inspected and recertified every 3 years and "reinspected" annually. A recertification is a complete inspection of the vessel to verify that all the applicable federal regulations are still met. In general, the scope of reinspections is the same as for a full inspection, varying in detail according to the conditions found at the reinspection. They focus on the vessel's equipment, watertight closures, operating practices, and accessible parts of the hull and machinery. Special attention is given to fire-fighting and lifesaving equipment. Parts of the hull and machinery that are prone to neglect and rapid deterioration are all examined. The inspector may conduct at his discretion a comprehensive inspection if conditions warrant.

As a vessel ages, maintenance and wear are increasingly important factors. Accordingly, the inspector must ascertain that all required fire-fighting, lifesaving, electrical, and mechanical equipment is on board and in satisfactory condition. If the vessel meets all the requirements (46 CFR 91.27), the COI remains in effect.

Underwater Body Inspections

Underwater bodies of vessels subject to inspection are also required to be thoroughly examined. The initial underwater body check, referred to as a drydock examination, is conducted in conjunction with the initial vessel inspection. Thereafter, the drydock examination is not directly linked to vessel inspection or reinspection, but may occur concurrently depending on the frequency established for underwater body inspections of different categories of vessels. For example, drydock examination is required every 18 months for 65-foot small passenger vessels (46 CFR 176.15). Because of the more severe operating environment, a more frequent underwater body examination might be appropriate for fishing industry vessels.

A drydock examination by the Coast Guard inspector includes:

- underwater hull and appendages;
- propeller(s);
- shafting;
- stern bearing;
- rudder;
- through hull fittings;
- sea valves; and
- external strainers.

In conducting the examination, the Coast Guard inspector also crawls through all compartments that include hull plates. The examination determines whether the structure is satisfactory: whether there are any holes, rotten wood, wasted fasteners, or other wasted internal components ("wasted" refers to the disappearance of metal due to corrosion). Inspectors are required to confirm that the vessel will remain seaworthy until its next required Coast Guard drydocking.

Safety Requirements

In order to be issued a COI, an inspected vessel must meet safety equipment requirements. It must have specified lifesaving and fire-fighting equipment, an emergency position-indicating radio beacon (EPIRB), pyrotechnics, line-throwing appliance, a fire control plan, and other safety items. All safety equipment must be Coast Guard-approved. Safety equipment that is not Coast Guard-approved may not be carried aboard inspected vessels.

The 1974 Safety of Life at Sea (SOLAS) Convention requires most commercial ships in ocean service engaged in international trade to be issued Cargo Ship Safety Construction and Cargo Ship Safety Equipment Certificates or a Passenger Ship Safety Certificate. These certificates state the numbers and types of lifesaving and other safety equipment aboard and are issued following a satisfactory inspection by a representative of the vessel's flag state (46 CFR 31.40, 71.75, 91.60, 176.35).

APPLYING VESSEL INSPECTION TO FISHING INDUSTRY VESSELS

The entire vessel inspection process could be employed for all fishing industry vessels or to the same degree for various categories or sizes of vessels in order to obtain some potential safety benefits.

Effectiveness of Vessel Inspection in Improving Safety

In determining whether to apply vessel inspection to fishing industry vessels, a key issue is whether inspection is effective in improving safety. If so, to what degree and at what cost? Studies of the effectiveness of the Coast Guard inspection program in reducing vessel-related casualties or fatalities or

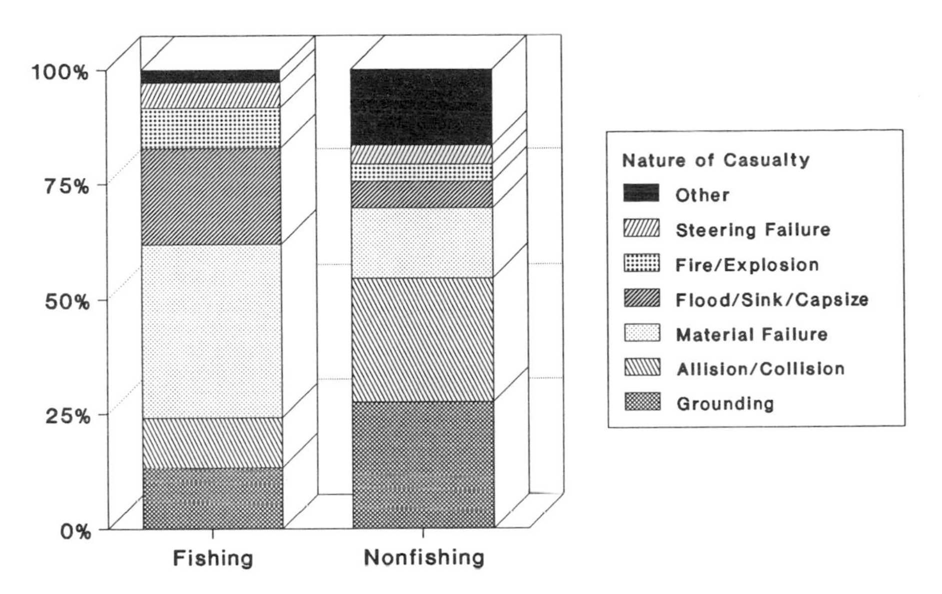

FIGURE G-1 Natures of vessel incidents. Source: USCG 1981-1987 CASMAIN data for fishing and nonfishing/nonbarge vessels.

in improving safety performance of vessels and equipment are not available; they also are not available on the benefits versus the costs of vessel inspection.

A rough indication of the relative safety performance of different segments of the commercial fleet may be obtained by comparing percentages of "involvements" (defined as participation of a vessel in a casualty) with the natures and causes of casualties (see USCG, 1990c). This approach assumes that exposure and usage factors are equivalent for the categories of vessels compared, does not generate casualty rates, does not by itself provide a measure of the overall material status of the fleet, and is not a direct measure of the effectiveness of inspection. Thus, the comparison that follows is at best a broad approximation.

The nature and cause data provided by the Coast Guard for comparing safety performance records consisted of CASMAIN data for fishing industry vessels and nonfishing/nonbarge vessels. The latter category includes inspected vessels (e.g., ships, commercial passenger vessels) and uninspected towing industry vessels. The data, summarized in Figures G-1 and G-2, indicate that material failure incidents and vessel factors as causes of casualties involving fishing vessels are twice as frequent as those involving other commercial vessels.

Including uninspected towing industry vessels in the data limit their utility for assessing whether comparable inspected vessels are more or less prone to vessel-related casualties than are fishing industry vessels. The nature of tug and barge operations results in significant exposure to allision/collision and grounding in restricted waters. The repair costs can easily reach CASMAIN

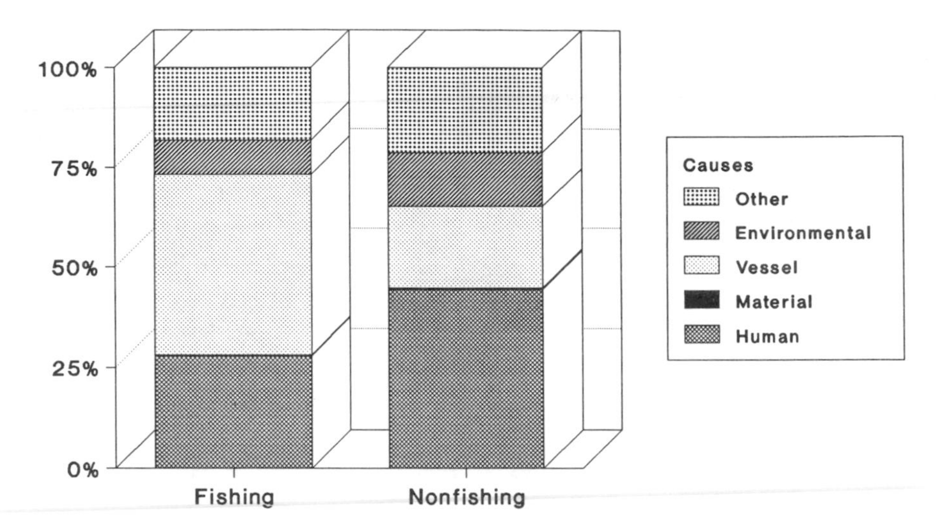

FIGURE G-2 Major causes of vessel incidents. Source: USCG 1981-1987 CASMAIN data for fishing and nonfishing/nonbarge vessels.

reporting thresholds. Thus, significant differences in these operations may bias the comparison.

What can be said is that operators of towing vessels are required to be licensed. This forces attention to safe operation of the vessel, including the physical condition of the vessel and equipment. Towing industry vessels are frequently units within corporate fleets. As such, they are more frequently operated under closer observation by management and Coast Guard marine safety officials than are fishing industry vessels. Unlike fishing vessels, towing industry vessels are usually operated under or in support of contractual arrangements rather than as a means of engaging in self-employment. This also tends to force attention to performing whatever maintenance is required to ensure that the vessel is available to meet its contractual obligations. All these factors motivate attention to maintenance.

Given the inconclusive findings from the data and analytical techniques and the absence of definitive research, the effectiveness of inspection as a safety-improvement alternative devolves to value judgments, collective opinions, and testimonial evidence. There are minimal material condition standards for each category of vessel subject to inspection. The nature of vessel inspection forces continuing attention to these standards. Generally, the conditions of vessels in other commercial sectors are somewhat better than they are in the fishing industry, considering the prevalence of material failures. The apparent better condition of inspected vessels is at least partly attributed to inspection and licensing requirements that compel operators to maintain certain equipment.

Which—inspection or licensing—contributes more to safety was not ascertained from the available data, nor was the relationship between benefits and costs.

Criteria for Establishing Which Fishing Industry Vessels Should Be Inspected

Safety performance, the nature of operations, regional operating conditions, and numbers of personnel on board could serve as criteria for determining or prioritizing which fishing industry vessels would benefit from inspection. The data reveal that conditions for which inspection could play a role exist across the entire national fishing industry fleet. It is evident that the potential consequences of marine casualties can vary significantly between vessels and operating environments. While the data are not conclusive as to which fisheries or activities are more prone to casualties, some vessels and fishermen are more likely to be involved in vessel-related casualties and fatalities than others.

Safety Performance Records

There is a common perception among fishermen that large, offshore vessels are where the safety problems lie and that vessels operating inshore do not need federal safety intervention, including various provisions of the CFIVSA. Findings that stand out are:

- Safety problems relating to material condition exist throughout the entire fishing fleet, on all sizes and types of vessels, on all fishing grounds
- Vessel-related casualties characterized as material failures are disproportionately high in the North Atlantic and West Coast regions (Figure G-3).
- Eighty-two percent of SAR events categorized as disabled and adrift involve failures of hulls, propulsion systems, or machinery (see Figure G-4).
- The largest number of vessel-related casualties and fatalities recorded in CASMAIN data involve vessels under 79 feet.
- Seventy-three percent of SAR cases in which property was lost or in danger of being lost involved vessels under 65 feet operating inshore and on inland waters.
- Very few vessel-related fatalities are recorded in CASMAIN involving commercial fishing vessels under 26 feet (see Appendix D).
- The average number of fatalities per vessel-related fatality incident is slightly higher for fishing industry vessels 65 feet or longer.
- Fishing industry vessels 79 feet or longer have significantly higher vessel casualty and fatality rates than those under 79 feet.

No single parameter that could be universally used as a threshold for imposing inspection stands out in the data. Rather, they indicate that a combination of parameters and different levels of inspection might be appropriate. Vessel

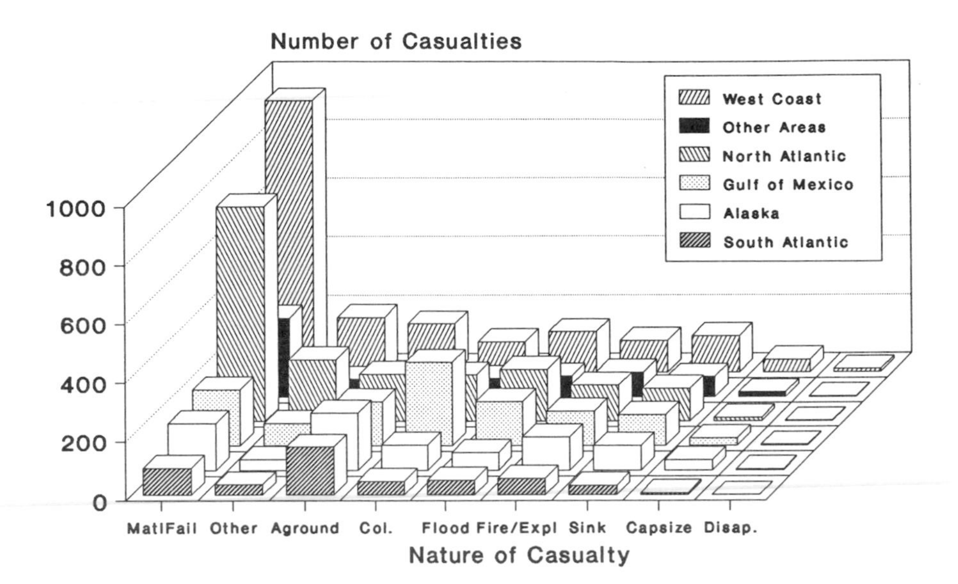

FIGURE G-3 Regional distribution of natures of incidents. Source: USCG 1982-1987 CASMAIN data for documented vessels.

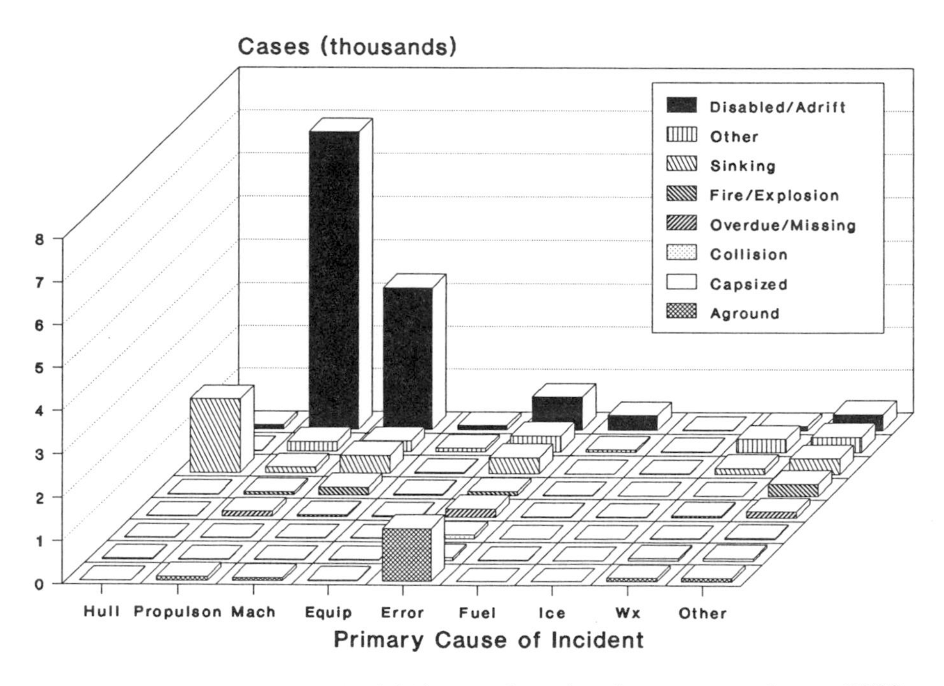

FIGURE G-4 Natures and causes of fishing vessel search and rescue cases. Source: USCG 1982-1987 SAR data.

attributes for imposing inspection options could include vessel safety performance records by type, construction materials, length, age, or other suitable measure.

Regional Factors

Significant regional variations in incidents involving vessels stand out in the CASMAIN data for documented vessels (Figure G-3). Material failure incidents are especially pronounced for the North Atlantic and West Coast regions. Contributing factors include operating environments, the vessel characteristics, and maintenance. While the data are not conclusive, the implication is that regional factors are important enough to be considered in framing vessel-improvement measures.

A vessel's operating environment can significantly affect the material condition of a vessel and its equipment. A significant number of fishing vessels of all sizes and most factory trawlers and fish processing and tendering vessels operate in Alaskan and adjacent North Pacific waters, although many are homeported in Washington and Oregon. These vessels are routinely exposed to harsh operating conditions in transit and on the fishing grounds that accelerate deterioration and create physical severe stress on the hull and equipment. Fishing vessels operating in the North Atlantic region and in Pacific Northwest waters similarly experience seasonally harsh conditions and deterioration.

A vessel's physical characteristics can also influence the types of incidents. The age of a large number of vessels in both the North Atlantic and West Coast regions is assessed as a contributing factor to disproportionately high numbers of material failure incidents—older vessels have higher rates of routine wear—in the aggregate data. Another contributing factor could be low profitability in some fisheries, which could adversely affect the ability to afford adequate equipment maintenance.

One would expect that material failure incidents would also be disproportionately high in Alaska; however, this is not indicated in the data. The reasons are not clear. Considerable attention has been directed toward safety in Alaska by Sea Grant, state educational organizations, and the fishing industry. Some vessels, particularly larger ones, have backup or redundant systems. Also, many of the vessels represent high capital investments; therefore, there are economic incentives to maintain vessel and equipment. On the other hand, the region's operating environment can lead to conditions that might quickly cause a minor problem to turn into a major incident, such as fire, flooding, or sinking.

Personnel Complements

Another factor meriting attention is the number of personnel aboard, primarily because of the potential for high loss of life. Factory trawlers, for example, have larger crews (including industrial workers), sometimes as high as 50-60. This is a new development for U.S. commercial fishing, which traditionally has been dominated by small and moderately sized vessels. Fish processing vessels, depending on size, may have more than 100 industrial workers aboard. The CASMAIN data for 1982-1987 indicated that many fatal incidents resulted in only one fatality and that the rate of incidence varied only nominally in relation to vessel length. When non-vessel-related fatalities were screened out, the number of fatalities per incident increased modestly relative to vessel length; however, the fatality rate increased dramatically (see Chapter 3).

During the study, a factory trawler with more than 30 persons aboard capsized in the Bering Sea during moderate weather while hauling in its trawl. The vessel and nine persons, including a National Marine Fisheries Service agent, were lost. Anecdotal information suggests that both vessel-related and human causes were associated with the disaster (Nalder, 1990; Matsen, 1990). Unless vessels are required to have load lines under 46 U.S.C.A. Chapter 51 or are subject to inspection, there are no minimum federal material condition standards or requirements that they must meet prior to operation. This will change for certain fishing industry vessels under the CFIVSA, as discussed earlier. The high estimated fatality rate for large fishing industry vessels from vessel-related casualties and the demonstrated potential for loss of life identify the size of the personnel complement as an important measure in determining whether to impose vessel inspection and to what degree.

Scope of Inspection

Conceptually, inspection could be applied to vessels, installed systems, and equipment. If inspection is selected as a safety-improvement alternative, what items to inspect and to what degree need to be determined. All primary and contributing causes of casualties need to be better identified, along with specific regional needs, in order to create a cost-effective approach for dealing with actual rather than perceived problems.

From a technical basis, the Coast Guard's existing inspection program is reasonably suited to large fishing industry vessels, which are comparable to inspected merchant vessels. Technical guidelines are available for design and construction of most fishing industry vessels, including NVIC 5-86 (USCG, 1986b) and the ABS *Guide for Building and Classing Fishing Vessels* (ABS, 1989), although additional guidelines are needed relevant to small fishing vessel stability. Various additional technical standards and guidelines could be adapted from those used by other fishing countries (see Appendix C) or developed especially for the fishing industry.

IMPLEMENTATION

The Coast Guard's vessel inspection program is resource intensive. Neither the Coast Guard nor industry overall appears to have the resources needed to implement a full-scale program of this type in the fishing industry. A less costly inspection alternative, such as self-inspection, marine surveys, or load lines, could be implemented more easily in the near term.

If the full Coast Guard vessel inspection process were expanded to include fishing industry vessels, the logical first step would be to bring the larger vessels under the program. This is being done de facto by the CFIVSA requirement for classification of new or converted fish processing vessels. Fishing industry vessels already subject to or otherwise involved with classification, marine surveys, and load lines may already meet some of the vessel inspection criteria, and the burden of formal vessel inspection could be less onerous on them.

Extending classification requirements or the full vessel inspection process to all fishing industry vessels 79 feet or longer could be predicated upon the estimated higher vessel casualty and fatality rates presented in Chapter 3 and the potential for high loss of life. However, the estimated rates are not statistically valid, and it cannot be said that they would apply equally to all large fishing industry vessels. Some vessels and fleets are better maintained than others. Because there are relatively few larger vessels, a modest research effort into fleet composition and vessel material condition could provide more insight on which larger vessels, if any, should be brought under the Coast Guard's vessel inspection program or a less rigorous inspection regime.

Fishing industry vessels under 79 feet present a different problem. Clearly, the largest number of breakdowns and vessel-related casualties and fatalities involve smaller vessels fishing relatively close to the coast. Generally, the smaller the vessel, the lower the casualty rate—although even the smaller vessel rates characterize commercial fishing as a dangerous occupation. But the data do not provide a sufficient resource for identifying and selectively applying a rigorous inspection program to those smaller vessels most prone to vessel-related problems or inadequate maintenance.

It appears that good material condition can be maintained. For example, self-insurance groups require meeting material condition and equipment standards. Thus, short of full vessel inspection, other measures hold promise of meaningful benefits at less cost to the industry and the government.

Potential Inspection Organizations

If federal inspection were required for the fishing industry, oversight would naturally devolve to the Coast Guard because of its longstanding leadership for this function. However, such a role would not necessarily mean that inspections should be performed by federal personnel. The nature and scope of vessel inspection would imply at which organizational levels such a program could be most effectively conducted. Who bears the costs of inspection is important but beyond the scope of this study.

The Coast Guard has a nationwide infrastructure, including field stations in or near most major fishing ports, which could be used to frame, oversee, or conduct an inspection program. However, the agency does not have the resources in reserve to establish or operate a full-scale vessel-inspection program for even documented fishing industry vessels. It has already established limited precedents that accept services provided by third parties, and the CFIVSA expands this approach. These precedents hold potential for more widespread usage in support of safety initiatives in the fishing industry. The Coast Guard advised the committee that some third-party interest has already been expressed.

The following noninclusive list of organizations potentially could be authorized to provide inspection services.

Marine Survey Organizations

A number of marine surveyors specialize in fishing industry vessels. This activity will also grow under the CFIVSA. As of August 1990, one marine surveyor organization and one marine survey company have expressed interest to the Coast Guard in an expanded third-party role in improving safety. Currently, however, the number of marine surveyors who specialize in fishing vessel surveys is insufficient to rapidly expand marine surveys to larger portions of the fishing industry (Expert, 1990).

Classification Societies

Classification societies such as ABS and Det norske Veritas are active in fishing vessel safety in the United States, and this role will increase under the CFIVSA. Extending classification requirements to more fishing industry vessels is one way to improve the fitness for service short of full inspection. However, the cost of building and maintaining a vessel to class standards is an important element that must be considered.

Insurance Organizations

The questionable viability of improving safety through active involvement of the insurance industry was presented in Chapter 7. However, it has been demonstrated that self-insurance groups can improve the physical condition of vessels by admitting only good risks and by applying more rigorous vessel and equipment safety standards than presently enforced by the Coast Guard. Expanding self-insurance groups may be possible in some areas, but has little potential for widespread application for various reasons, not the least of which is the marginal profitability of many existing fishing operations.

State Governments

The potential for state government administration of vessel inspection was introduced in Chapter 4. For example, extension of motor vehicle safety inspection programs to state-numbered fishing vessels is an intriguing concept. Potential infrastructures already exist in some cases and could be adapted for vessel safety inspections. Such a program could be self-supporting through inspection fees. However, there is an important distinction relevant to inspection as practiced by the Coast Guard. State automobile inspections are analogous to Coast Guard compliance activity in that they check—albeit on a regularly scheduled basis—that a vehicle meets all applicable safety (and pollution) laws and regulations. The states do not inspect vehicles during construction, however. Thus, if inspection were to be administered by the states, it would most likely take the form of a shore-based compliance examination program. State boating administrators focus their attention toward the recreational boating public (an estimated 17 million recreational vessels) and rely on a pass-back of federal funds to support many of their activities. Most states have been reluctant to impose recreational vessel operator training or licensing, although a few are considering this action on a selective basis, principally for young operators. There is no indication that the states would willingly undertake a safety program for fishing vessels or that such programs would be undertaken without funding support from the federal government.

Industry Organizations or Associations

Various fishing industry organizations have actively promoted safety, conducted training sessions, provided videotapes on inspection and maintenance, and engaged in similar activities. None is known to provide services akin to inspection, although safety materials from industry sources provide guidance on vessel maintenance. It does not appear that any industry organization is structured to undertake national or regional administration of an inspection program. However, drawing on the experience of self-insurance groups, it is potentially feasible for owners' associations to establish and administer safety inspection programs. If accepted by the Coast Guard, they could provide an alternative to direct federal involvement.

Owners or Operators

Some owners and operators provide or conduct substantial programs to ensure the physical condition and safety of their vessels. The Coast Guard potentially could accredit existing programs of this type that meet whatever criteria might be imposed in lieu of federal inspections. This option is potentially most applicable for fleet owners.

If self-inspection were adopted to motivate attention to vessel and equipment condition, administrative responsibility would fall on vessel owners, regardless of vessel size. Whether to conduct such an inspection with vessel personnel or engage a marine surveyor would be the responsibility of the owner and operator.

APPENDIX

H
Safety Considerations on a Longline Vessel

Captain Bob Jacobson

Solicited Expert Accounting for West Coast Commercial Fishing Vessel Safety Assessment
February 1990

Let me preface my remarks by stating that my hook-and-line longlining experience has been limited to two species—sablefish (black cod) and halibut. Most of my fishing ventures for those two species have taken place in the Bering Sea or Gulf of Alaska on an 84-foot vessel. While the following remarks are about my experience in Alaska, they directly relate to safety issues experienced by longline fishermen for those two species off Oregon, Washington, and California, since the gear and methods are very similar.

First let's talk about gear. It is comparable for both species. Baited hooks are attached to a nylon or poly groundline that is "set" over the stern of the vessel. Hook spacing intervals may range from 3 feet for black cod to as much as 30+ feet for halibut in certain areas. The groundline is anchored on the ocean bottom at each end, and also buoyed on the surface. Individual sets may range in length from under a mile to 4-5 miles.

I'll now attempt to outline the safety-related issues by fishery.

HALIBUT

Management of the halibut fishery has changed dramatically over the past 10 years. A tremendous increase in both the number of vessels operating in this fishery, and the efficiency of those vessels, has forced the management authority, the International Pacific Halibut Commission, to drastically reduce the length of the fishing season. It wasn't too many years ago when commercial halibut fishermen in area 3A (Gulf of Alaska) had well over 100 days a year fishing time. By 1989, the length of that season had been reduced to 4 days (one 24-hour opening in each of the months of May, June, September, and October). This management change has obviously had an impact on the fishing strategies employed by industry members. The same season-shortening process has occurred in area 2A (Oregon, Washington, California), where fishing season length has been reduced from several weeks to several days.

We employ a crew of seven (including the captain) for fishing halibut. It is likely that we will have at least one, and perhaps as many as three, new crewmen for each halibut opening. By this I mean that they are new to the boat, not necessarily new to the fishery. All crewmen are required to sign a crew employment form that we have developed. There is no standardized procedure in this industry for identifying available crewmen. Our first option is to hire an individual whom one of us knows personally to be experienced. If that fails we go to a list of names compiled from those who have shown the initiative to call or stop by the boat to look for a job.

Remember that with the very short openings that now face this industry, there's not much time for OJT (on-the-job training). It follows that the more inexperienced crew we are forced to hire for an opening, the more inefficient, less productive, and potentially dangerous our fishing operation will be. I should indicate at this point that while there are dangers involved in any fishery, it is my opinion that the longline fisheries for halibut and sablefish pose fewer personal injury risks than do other fisheries in which I've been involved. In the 8 years I've participated in the longline fishery, we've never had a major accident. We have had a very few minor knife cuts and puncture wounds.

We attempt to fish between 12,000 and 15,000 hooks during a 24-hour opening. The hooks are spaced at 9-foot intervals and are permanently attached to the groundline. This is commonly referred to as "stuck" gear. The groundline with hooks attached are stored in plastic tubs, with each tub holding approximately 1,200 feet of groundline and 140 hooks. We have a little over 100 tubs of gear on the boat.

Many vessels fish "snap-on" gear, where the hooks are not permanently attached to the groundline, but are snapped on with a safety-pin-type snap when the gear is set and snapped off when the gear is hauled aboard. Snap-on gear is used by a majority of the smaller vessels participating in the fishery, since it can be stored more compactly and therefore takes up less deck space.

The crew cuts all the bait and baits all 15,000 hooks prior to the time the boat leaves port for a 24-hour opening. This activity poses no particular safety hazard other than the threat of cutting a finger with the knives used to cut the bait. We time our departure from port so that we arrive on the grounds several hours ahead of the season opening. When fishing in the Kodiak, Alaska, area, our run to the grounds may take from 8 to 40 hours, depending on the area fished.

Prior to the departure we attempt to talk with all the new crew about safety equipment—where the survival suits are stored, how to put them on, and how to deploy the two inflatable rafts—and about their responsibilities as they relate to the overall fishing operation.

Generally speaking, the area from the harbor to the open ocean poses a greater navigational risk than the open ocean itself. Congestion from increased numbers of vessels operating in a confined area, plus the need to navigate through buoys and reefs when leaving or entering port, contribute to this risk. Only the captain or alternate captain on our vessel is allowed to be on wheelwatch when we are leaving or entering port.

On the run to or from the grounds, wheelwatches are shared among the experienced crew. We attempt to break in an inexperienced crewman gradually to this responsibility by making him accompany either the captain or alternate captain on one or several of his watches. During this indoctrination, we attempt to introduce the crewman to all of the navigational equipment, steering system, and vessel alarm system, including the watch alarm. Generally, after several indoctrination periods, we'll put the new man on watch during daylight hours, when others of the crew are awake, then on to nighttime watches when he (and we) feel comfortable that he's ready. We continually try to stress two issues: (1) that when he's on watch he's responsible for the safe passage of the crew and the vessel, and (2) that going to sleep on watch is grounds for dismissal after returning to port. The watch alarm is set and used by all those on wheelwatch.

All crew are awakened at least 10 hours prior to the season opening. After eating, it's on to the back deck where final preparations are made. With only 24 hours to fish, there's no time for equipment breakdowns, poorly tied knots, or being anything less than fully prepared.

There's also no room for injuries, and that's a point we continually stress to all the crewmen, but particularly to the new men—"Proceed with the responsibilities that we've discussed with you, but proceed cautiously until you've gained a better understanding of the processes and the dangers involved."

Setting the gear generally takes from 3 to 4 hours, the length of time depending on several variables. Experienced crew always do the setting. The baited gear in tubs is placed on a table near the stern of the boat where the end of the line in one tub is tied to the end of the line in the next tub until the lines in all tubs that we are going to use in that set are tied together. Tying the lines together is a job only for the experienced. One bad knot can ruin your day. Buoys and a flagstick are the first into the water. Flagsticks are generally made of either aluminum or calcutta and are long and bulky. Care must be taken when handling the flagsticks on board that another crewman standing nearby doesn't get hit, or poked in the eye.

Attached to the flagstick/buoys is the buoyline. The length of the buoyline depends on the depth being fished. No particular safety problems to this point. Next overboard is the 35-60-pound anchor, followed by the groundline with the baited hooks attached. When setting groundline with baited hooks permanently attached, a "setting chute" is used. The tub from which the groundline is being set is placed in a position under the chute so that the groundline and hooks leave the tub vertically, passing over the top of the chute and overboard. Once all the groundline and baited hooks from one tub are over the stern, that tub is quickly removed and the next tub is quickly slid into position. That process is repeated until all the tubs on that set are emptied. An anchor is attached, followed by the buoyline and flagstick.

The gear setting can be dangerous. Hooks are leaving the tub at the rate of approximately one per second, or faster. A crewman hooked during this process may be dragged overboard, or at the minimum, sustain some hooking-induced injury. We have several sharp knives located in sheathes next to the setting chute for emergency use to cut someone free who might accidentally get hooked. We've never had to use them. I can recall, however, several occasions where the hooks from one tub tangled with the hooks from the next tub, creating a potentially dangerous situation where all hands were scrambling to stay clear of the flying hooks. My instructions to the crew are always to stand well clear when any tangles occur. Don't try to reach in to unsnarl it!

During the setting process, the boat is proceeding full speed ahead with the captain in the wheelhouse. We've found it very useful to have an intercom system from the wheelhouse to the setting area on the back deck so that the captain can communicate directly with the crew when a problem arises.

Snap-on gear is set differently. The groundline is generally stored on a large, hydraulically powered reel. After the buoy, flagstick, buoyline, and anchor are set, the crew manually snap each hook and gangion (leader) onto the groundline as it passes over the stern of the vessel. I have fished both snap-on and stuck gear, and would probably conclude that when setting the gear, there may be more risk of injury with the snap-on gear, since each snap and hook is handled individually. But that's really a judgment call.

Once all of our sets are made and all of the gear is in the water, we return to the first set we made and begin the retrieval process, knowing that we've only got 19-20 hours to complete the task. The gear is retrieved over the starboard side of the vessel just aft of the wheelhouse. The flagstick is first aboard, followed by the buoys, buoyline, and anchor. A hydraulically powered line hauler mounted directly across the deck near the port rail is used to pull this gear, as well as the groundline that follows. The crew responsibilities during the retrieval process are as follows: one experienced man (generally the captain or alternate captain) gaffing the fish as they come aboard—he also runs the vessel using throttle and steering controls mounted next to his gaffing station (yes, that means that there is no one in the wheelhouse when we are running gear); one experienced man at the line hauler coiling the hooks and groundline into the empty tubs; one man unhooking the fish and putting them on the table for cleaning; and four men dressing fish (one man dressing fish is also responsible for carrying full tubs of gear from the hauler back to the gear storage area and carrying empty tubs back to the man operating the line hauler).

Halibut must be 32 inches in length to be legal and may weigh from 15 to 400 pounds. All fish are landed using gaff hooks. The person doing the gaffing (generally referred to as the "roller" man) assumes a major responsibility in our operation. Not only does he gaff all the fish, but he must also run the vessel in a safe and efficient manner, always being aware of other vessel traffic in the area. In addition, he's the one person in the crew who is most exposed to the weather and sea conditions. There have been times, when fishing in tough weather, that I've been completely submerged under a wall of water when a wave breaks over the side of the vessel.

As the groundline is retrieved, each fish is gaffed in the head and lifted aboard. An innovation that a few vessels, including ours, now use is to cut away about a 3-foot-wide section of the 4-foot-high bulwarks so that the man doing the gaffing doesn't have to physically lift the big fish over the bulwarks onto the deck, but rather only has to lift them to deck level, where they slide through the opening in the bulwarks. While this means more water on deck (not a safety issue in this case), it also means considerably less physical strain on the roller man and certainly less chance for back or shoulder injuries. When a particularly large fish appears alongside the boat, it may take as many as four or five men, each wielding a gaff hook, to bring him aboard. This is certainly a potentially dangerous activity which could result in a severe puncture wound to one of the crew if a gaff hook slipped out of the fish. There's also a chance that someone might fall overboard during this endeavor, since it involves leaning over the bulwarks to get your gaff hook into a wildly thrashing, 200- to 300-pound fish.

The groundline retrieval process continues until all hooks in that set are in the boat. The same process continues until all sets have been run and the season closes at the end of the 24-hour open fishing period.

A couple of other potentially dangerous conditions may affect the safety of the roller man. There are times when the groundline snags on the ocean bottom and becomes extremely tight. If the groundline breaks at a point between the hauler and the roller man, it can snap back overboard, hooking the roller man on the way over. Or, if the line jumps out of the hauler under tight line conditions, it can snap back, hook the man at the roller, and either pull him overboard or inflict a hooking-related injury. As we do in the gear-setting area, we keep a couple of sharp knives handy, that the roller man might get to in case of an emergency.

As I indicated earlier, we generally have three to four crewmen at the cleaning table at all times. Inexperienced crew are generally assigned to this task. Many have had no previous experience cleaning any kind of a fish, let alone a giant halibut. Just wrestling them up onto the cleaning table can be a multiperson, energy-draining, backbreaking task, particularly in rough weather. Experienced crew will provide instructions for the first couple of hours, then they are on their own, with instructions that it's better to take their time and be safe than to have a knife-related accident that could immediately end the trip for all of us.

I mentioned earlier that we cut out a section of our bulwarks to ease the strain on the roller man in getting the fish aboard. We also installed an inclined ramp that extends from the cut-out bulwarks upwards to the cleaning table so that after the hooked fish comes aboard, but before it's unhooked, it's pulled up the ramp and onto the cleaning table by the hydraulic line hauler. During times of heavy fishing the cleaning crew simply can't keep up, so the cleaning table becomes filled with fish waiting to be dressed, and uncleaned fish end up on the deck where they must be physically lifted onto the table. Removing the bulwarks and installing the ramp to get big fish on the table were positive steps from both a safety and efficiency point of view.

As you're undoubtedly aware, halibut are a flat fish and are very slippery. While we use 4 × 24-inch deck checkers to keep the uncleaned fish on deck from sliding back and forth in rough weather, there have been occasions when our crew got so far behind cleaning that we were forced to temporarily halt our fishing operations and have all hands clean fish. Large amounts of halibut on deck, in rough weather, may create vessel stability problems, since water on deck does not clear properly through the scuppers as it should.

Most smaller vessels send one or two crewmen into the fish hold to hand-ice fish after they are cleaned. Working in the cramped confinement of a fish hold with fish of this size is an arduous task, particularly in bad weather. On our vessel we are able to flood the fish holds. Just prior to leaving port we take 30-40 tons of ice and flood the hold, creating a slurry of ice and salt water. When the fish is cleaned, it's simply dropped through a hole in the cleaning table into the slush ice in the hold. In my opinion, this is certainly a safer and more efficient way of refrigerating the product. Care must be taken not to allow "slack" water in the tank when using this system. Potential stability problems may arise if that occurs.

The crewman carrying tubs full of recently run gear from the line hauler back to the gear storage area has one of the most physically demanding jobs on the boat. He must maintain his balance while walking on a slippery wooden deck carrying a tub full of water-soaked groundline, hooks, and unused bait that may weigh up to 100 pounds. He may, at times, be walking over fish lying on the deck. While it's only 25-30 feet, he may make that trip 100 times in a 24-hour opening.

Like most of the larger halibut boats, we have an aluminum "shelter deck" that covers about three-quarters of the back deck. As the name implies, the shelter deck provides shelter for the crew from weather and sea conditions, thereby making their jobs a little more comfortable and certainly safer, particularly when fishing in tough weather and sea conditions.

I haven't said much about weather to this point. It is a factor for all vessels involved in this fishery, but becomes less of a consideration as vessel size increases. Look at it this way—with only a limited number of preestablished days to fish, it's important that we optimize our production during those days. There's no question that, at times, we fish in weather and sea conditions that we may not otherwise be fishing in if there was another alternative. I can recall on several occasions continuing to fish in 60-mph winds and 25-foot seas, which wasn't particularly comfortable, We confront such conditions by slowing down our operation in an attempt to minimize risk of injury to crew.

Once the last fish is aboard we begin the return trip to port. All crew have now gone a minimum of 26 hours without sleep, with the exception of one experienced man who's been in the bunk for 2 or 3 hours. He'll get up and take the first wheelwatch. Any fish that haven't been cleaned are taken care of. The boat is cleaned up and fishing gear properly stowed. Fatigue has set in with all of us, but that's just a way of life in the fishing industry.

The experienced crew will rotate wheelwatches, with the captain or alternate captain assuming that job prior to entering the congested harbor area. Once we arrive at the processing plant, we wait our turn to unload. Our crew does the unloading. There's always the potential here for having a brailer full of fish break on the way out of the fish hold and fall back into the hold on the unloaders. They are instructed, however, to stand well clear of the fish leaving the fish hold. Hopefully, we've got a good trip, but there are never any guarantees in this industry.

SABLEFISH

The sablefish longline fishery has not been as tightly regulated in recent years as has the halibut fishery. Season openings continue to be long enough in most areas that vessels fish repeated trips of several days' duration instead of the periodic 1-day trips common to the halibut fishery.

While the gear and methods in this fishery are very comparable to halibut, there are differences in the operation that may have safety-related consequences. A brief discussion of each of those issues follows:

1. Most of our sablefish trips are 6-7 days in length. As a result, there is more time for OJT for inexperienced crew. In my opinion that's an important factor in minimizing the chance of personal injuries to the inexperienced crew. They are not under the pressure, either perceived or real, to perform immediately as they are in the halibut fishery.

2. In my opinion, the crew fatigue factor is greater in the sablefish fishery than for halibut. The crew averages 4-5 hours of sleep a night during the 6- or 7-day trip. The experienced crew has learned, over the years, to adjust to this rigorous schedule. Invariably, the inexperienced are accustomed to more sleep and, as a result, may have a difficult time adjusting. As we all know, accidents tend to happen more frequently after fatigue sets in.

3. Longer seasons mean that the need to fish under adverse weather and sea conditions is minimized. As a result, when the weather deteriorates while we're on a sablefish trip, we will cease fishing and either jog into the weather until it improves, or will run as much as 8-10 hours to get to a safe anchorage. When we're jogging in rough weather, an experienced man is always on the wheel. If we choose to anchor, once again the experienced crew assumes all the responsibilities of running the vessel and dropping the anchor.

4. Sablefish are much smaller in size than halibut. Most of the fish are in the 4- to 6-pound category, with a few weighing as much as 15 pounds. The potential of lifting-related injuries occurring in this fishery is certainly less than in the halibut fishery.

5. As mentioned earlier, the gear and methods are similar in the two fisheries. As in the halibut fishery, minor knife cuts, gaff hook, and hook punctures are the most common personal injuries that occur.

APPENDIX

I
Survival and Emergency Equipment: Technical Analysis

This appendix provides a technical analysis of basic survival and emergency equipment issues as they pertain to the commercial fishing industry. The analysis is based principally on interviews with Coast Guard and Underwriter Laboratories (UL) staff with functional responsibilities for safety and survival equipment.

EQUIPMENT REQUIREMENTS

Survival equipment required by federal regulations prior to July 1990 varied by size of vessel, operating area, nature of employment or use, and personnel complement of vessel. Table I-1 summarizes safety and survival equipment required aboard uninspected fishing vessels. At a minimum, all uninspected fishing vessels must carry one readily accessible, wearable personal flotation device (PFD) of the appropriate type and size for each person on board. A prescribed number of fire extinguishers and certain other equipment must also be carried. Visual distress signals are not required. All equipment must be installed, accessible, or worn, as appropriate, according to Coast Guard regulations and specifications for approved equipment. Lifesaving devices that are not Coast Guard-approved may be carried aboard vessels to augment that which is required. Life rafts and VHF-FM portable radios are not required by the Coast Guard as survival equipment but may be carried aboard uninspected fishing vessels as additional gear (see McCay et al., 1989; National Transportation Safety Board [NTSB], 1987). It is also noted that basic safety equipment items such as protective headwear, flexible wire mesh gloves, and high-traction

TABLE I-1 Federal Safety and Survival Equipment Requirements for Uninspected Fishing Vessels in Effect July 1990

Required Survival Equipment[1]	Under 26 Ft	26-39 Ft	40 Ft & Longer
Life Preservers/PFDs[2]	Type I/II/III	Type I/II/III	Type I
Type V Antiexposure Coveralls[3]	No	No	No
Exposure (Survival, Immersion) Suit[2]	No	No	No
PFD Light[4]	Yes	Yes	Yes
Type V Wet Suit (Diver Style)[3]	No	No	No
Life Rafts	No	No	No
Lifeboats	No	No	No
Ring Life Buoy	No	Yes	Yes
Life Floats	No	No	No
EPIRBs[5]	Yes	Yes	Yes
Visual Distress Signals[6]	No	No	No
VHF-FM radios[7]	No	No	No
Fire Extinguisher	Yes	Yes	Yes
First Aid Kit	No	No	No

Notes:

[1]Consult pertinent 33 and 46 CFR subchapters and product labels for specific federal requirements, product descriptions, and special provisions and instructions for usage.
[2]An exposure suit or *commercial* Hybrid Type V PFD meeting criteria specified in 46 CFR Part 25 *for commercial vessels* may be substituted for a required life preserver, buoyant vest, or marine buoyant device.
[3]Not required but may, if worn, be used to satisfy the requirement for Type I/II/III devices aboard uninspected vessels under 40 feet long.
[4]Required for ocean, coastwise, and Great Lakes voyages.
[5]A 406-MHz EPIRB with automatic activating devices is required aboard all uninspected fishing vessels on or after May 17, 1990, when these vessels operate on the high seas (generally considered as outside of 3 nautical miles from shore). Within certain constraints specified by the Federal Communications Commission, older Type A EPIRBs are grandfathered until August 1, 1991. Vessels without galleys or berthing facilities were exempt pending action on a proposed rule allowing a manual in lieu of an automatic device.
[6]Required for recreational vessels but not for uninspected commercial vessels.
[7]There is no requirement for portable VHF-FM radios as safety or survival equipment.

waterproof footwear are not required aboard uninspected fishing vessels by federal regulations and are not in widespread voluntary use (Nixon and Fairfield, 1986).

Ongoing Coast Guard rulemaking under authority and direction of the Commercial Fishing Industry Vessel Safety Act of 1988 (CFIVSA) will add or expand requirements for survival equipment including life rafts and, above a prescribed latitude, immersion suits (Federal Register, 1990). During a meeting of the Commercial Fishing Industry Vessel Advisory Committee (CFIVAC) in Seattle, Washington, during October 1989, it was stated that some manufacturers had not achieved favorable economic returns when they surged to produce immersion suits to meet earlier Coast Guard rulemaking requiring them aboard certain inspected vessels (CFIVAC, 1989). Manufacturers were reported to have no economic incentive for surging production to meet similar requirements that may be applied to the fishing industry. Regardless of the validity of the statements before CFIVAC, phasing of new equipment requirements must be consistent with manufacturing capability.

Coast Guard-Approved Equipment

Formal approval normally requires manufacture to Coast Guard standards and testing by accepted independent laboratories, although the Coast Guard has occasionally approved equipment in advance of formal standards in order to meet special needs. Approvals typically take several months, although several years is sometimes required to complete the process if a manufacturer experiences difficulty in meeting requirements. To satisfy Coast Guard equipment requirements, there is no intermediate Coast Guard designation analogous to "patent pending." Survival equipment is either Coast Guard-approved or not approved.

Coast Guard construction and performance standards applicable to approval of prototype safety and survival equipment are extensive and exacting. It is important to note that for PFDs, the Coast Guard emphasizes safety of the recreational boating public, which comprises the principal market segment for this equipment. The agency attempts to minimize the opportunity for recreational boaters to inadvertently defeat equipment design. For example, flotation material must be fixed to the device rather than removable. This policy does not preclude the Coast Guard from approving two-piece systems as long as each component meets all other applicable criteria. The same standards also apply to basic personal flotation equipment that is required aboard most uninspected fishing vessels. Promising policy and technological advances in Canada have resulted in adoption of standards for special-purpose, two-piece work suit systems specifically designed for use aboard Canadian commercial fishing vessels. This standard potentially could be adapted for use in the United States. A summary of Canadian fishing industry safety activities is presented in Appendix C.

Use of Equipment Not Approved by the Coast Guard

Equipment for which a manufacturer has requested but not received Coast Guard approval may not be carried aboard a vessel in place of required Coast Guard-approved equipment. Likewise, the Coast Guard does not accept as approved any item acquired for use before it grants an approval to a product and the appropriate label is affixed by the manufacturer as an element of product construction.

Catalogs offering Coast Guard-approved equipment are required by Coast Guard and UL guidelines to be consistent with use constraints. However, manufacturers are not prohibited from advertising that equipment has been submitted for approval, a common practice in catalogs and promotional material. The Coast Guard has occasionally intervened with some manufacturers to motivate better catalog representation of approval status and usage criteria for lifesaving equipment. Use of nonapproved equipment is permitted aboard uninspected fishing vessels as long as the required Coast Guard-approved equipment is also carried as prescribed by regulations. Performance of nonapproved equipment varies greatly (see Castle, 1988; National Transportation Safety Board, 1989a). Coast Guard technical personnel report that effectiveness may vary among comparable types of Coast Guard-approved equipment, but each item must fully satisfy the agency's standards.

Equipment Costs

Survival equipment costs range from reasonable to expensive. Actual costs depend on equipment type and nature. Basic PFDs, fire extinguishers, and visual distress signals are inexpensive, whereas a small, covered life raft or emergency position-indicating radio beacon (EPIRB) can cost well over $1,000. For example, extremely portable, lightweight, Coast Guard-approved, covered life rafts ("valise packs"), about the size of a large backpack, are available for use as optional lifesaving equipment on small, uninspected vessels. This equipment is expensive, typically $2,400 to $5,000.

EQUIPMENT AVAILABILITY

A wide variety of lifesaving equipment is available for use by operators and crews of uninspected fishing vessels. Table I-2 lists general categories of survival and emergency equipment and availability of Coast Guard-approved equipment in each category. In some cases, various items within categories bear dual Coast Guard approvals to satisfy carriage requirements for both recreational and commercial vessels. Applicable regulations and labeling on each item must be consulted to determine whether it satisfies regulations for the vessel on which it is used.

Personal Flotation Devices

The five types of Coast Guard-approved PFDs are summarized in Table I-3. Nearly all Coast Guard-approved PFDs, including work or deck suits (antiexposure coveralls) and immersion/exposure (survival) suits, require inherent, permanently fixed flotation. Inherent flotation material adds bulk. Some manufacturers have reduced impacts to mobility through design modifications. For example, Type III devices are usually designed for convenience, such as the well-known "float" coats. Under Coast Guard standards, approved Type III PFDs are not required to hold the face of an unconscious wearer clear of

TABLE I-2 Safety and Survival Equipment Availability

Type of Equipment	CG-Approved Available	Non-CG-Approved Available
Personal Flotation Devices (PFDs)[1,2]	Yes	Yes
Antiexposure Coverall (Deck Suit, Work Suit)	Yes	Yes
Immersion Suit (Exposure Suit, Survival Suit)	Yes	Yes
Hybrid PFDs[3]	No	Yes
Inflatable PFDs[4]	No	Yes
Wet Suit[5]	No	Yes
Ring Buoys	Yes	Yes
Life Floats	Yes[6]	Yes
Life Rafts	Yes	Yes
EPIRBs	Yes[6]	No
Visual Distress Signals	Yes	Yes
VHF-FM Radios	No	Yes
Fire Extinguisher	Yes	Yes
First Aid Kit	No	Yes

Notes:

[1]Available in a wide variety of types and styles.
[2]Coast Guard (CG)-approved immersion suits and Type V lifesaving devices are not available for small children and infants. Type III lifesaving devices are available for children 30 pounds in weight or greater.
[3]Hybrid PFDs combine inherent flotation and air chambers. Coast Guard-approved hybrids are available for use aboard recreational vessels. No hybrid types have been submitted for Coast Guard approval for use aboard commercial vessels.
[4]No inflatable lifesaving devices with dual air chambers and automatic inflating devices have been submitted for Coast Guard approval for use aboard commercial vessels. The Coast Guard does not approve inflatable PFDs for recreational use.
[5]No wet suits have been submitted for Coast Guard approval for use aboard commercial vessels. One diver-style wet suit has been approved as a Type V PFD for recreational use by water skiers.
[6]Self-activating EPIRBs are available and became mandatory for some uninspected fishing vessels operating on ocean waters on May 17, 1990. EPIRBs and VHF-FM radios are type approved by the Federal Communications Commission. It is illegal to use nonapproved radio devices on U.S. vessels.

the water, nor are they intended to support extended survival in rough water. Hybrid PFDs combine inherent flotation and air chambers, reducing overall bulk and minimizing impacts on mobility. Hybrids have been approved as Type V special-purpose PFDs for recreational use. These hybrids are often advertised as Coast Guard-approved Type V PFDs without reference to approval for recreational use only. Similar equipment is feasible under existing Coast Guard standards for commercial use. However, no equipment of this latter type has been submitted to the Coast Guard for approval. Any hybrid PFD subsequently approved by the Coast Guard for commercial use would also be identified as a Type V special-purpose PFD (as are approved work vests and antiexposure coveralls). Potential buyers and users must read the entire manufacturer's label,

TABLE I-3 Coast Guard Personal Flotation Device Definitions

Offshore Lifejacket (Type I PFD) Designed for use in open, rough, or remote water where rescue may be slow coming. Required to turn most unconscious wearers face-up in the water. Must be of highly visible color.
Nearshore Buoyant Vest (Type II PFD) Designed for use in calm, inland water or where there is a good chance of fast rescue. Required to turn many unconscious wearers face-up in the water. Not intended for long hours in rough water.
Flotation Aid (Type III PFD) Designed for calm, inland water, or where there is a good chance of fast rescue. Not designed for extended survival in rough water and will not hold the face of an unconscious wearer clear of the water. Generally, the most comfortable PFD for continuous wear. Available in many styles, including vests and flotation coats.
Throwable Device (Type IV PFD) Designed for calm, inland water with heavy boat traffic, where help is always nearby. Not intended for unconscious persons, nonswimmers, or for many hours in rough water.
Special-Purpose Devices (Type V PFD) Approved only for special uses or conditions. Must be used in accordance with limits specified on the label.
Type V Hybrid Device Combines inherent flotation and inflatable bladders. Has high flotation when inflated. May not adequately float some wearers unless partially inflated. Equal to either Type I, II, or III performance as noted on the label, but must be worn and used as specified on the label in order to be counted as a regulation PFD.

which states the scope of approval, to determine whether individual items would satisfy Coast Guard criteria for use aboard uninspected commercial vessels.

Coast Guard-Approved Inflatable PFDs

Although possible under Coast Guard regulations, the only inflatable PFDs approved to date *for commercial application* are for use aboard commercial passenger vessels with limited stowage space (e.g., passenger submersibles). The Coast Guard standard requires dual air chambers and automatic inflation among other extensive requirements needed to obtain approval. A fully functional automatic inflation device at all water temperatures is estimated to cost as much as $120. A device of this type is used as a basic component of U.S. Navy inflatable PFDs (to which Coast Guard regulations do not apply). The Coast Guard does not approve fully inflatable PFDs for recreational use because of maintenance considerations.

Infant- and Child-Sized Immersion Suits

There are no Coast Guard-approved PFDs or immersion suits that provide hypothermia protection for infants and small children. A number of fishermen from coastal Alaskan waters, in response to the Coast Guard's Notice of Proposed Rulemaking under the CFIVSA, petitioned the Coast Guard to encourage the manufacture, sale, and approval of such equipment. The need was attributed to family fishing operations in Alaskan coastal waters where—by choice or economic necessity—the boat doubles as workplace and home (see Page, 1989). The Coast Guard corresponded with various manufacturers concerning this matter. Some were not interested because of the small market segment and product liability, but several manufacturers developed and tested infant/small child immersion suits in Alaska with mixed results. The Coast Guard has received no proposals from manufacturers requesting approval of protective clothing of this type. However, the Coast Guard has advised manufacturers that it will, on a case-by-case basis, consider approving immersion suits in sizes other than the three prescribed by regulation.

PFD Lights

PFD lights are required for uninspected fishing vessels operating on ocean waters, coastal waters, and the Great Lakes. The PFD light must be fully operable and properly attached in order for it to count as required equipment during a Coast Guard boarding. Because they are not required to float, if the light is not properly attached to the PFD, it may sink. A wide variety of designs and battery types are used by manufacturers with variations in performance. Some lights require manual activation; others are water-activated. Marker lights that float are available with either manual or automatic activation devices.

EPIRBs

EPIRBs are used by offshore or distant-water fishing vessels throughout the industry. Of nearly 40,000 first alerts from EPIRB signals detected by satellites during calendar year 1989, 92 percent were not heard a second time. Furthermore, the source of only 2 percent of the total number of alerts was determined, and only 35 were verified as distress cases. The poor alert record is directly associated with use of 121.5 MHz as the transmitting frequency for older classes of EPIRBs. This frequency is shared with the aviation community, and an undetermined number of false alerts undoubtedly came from those units. The older EPIRB beacon is inadequate for satellite position fixing. In view of these factors, the Coast Guard does not consider a first alert on 121.5 MHz as sufficient justification to initiate a search endangering rescue aircraft and personnel (Embler, undated; Lemon, 1990a,b; Pawlowski, 1987; unpublished Coast Guard statistics, 1990).

A vivid demonstration of the marginal effectiveness of 121.5-MHz EPIRBs occurred during March 1990. A 60-foot steel dragger transiting from Washington State to Prince William Sound, Alaska, with three persons aboard was reported overdue after departing from Cape Spencer, Alaska, and a search began. The vessel was equipped with a life raft, 121.5-MHz EPIRB, and survival suits. Coast Guard rescue aircraft located the vessel's life raft and attached EPIRB on the beach near Cape Spencer. The life raft was sighted visually before a weak signal was detected from the EPIRB. On-scene personnel recovered the gear and determined that the life raft had self-deployed and the EPIRB self-activated. The survival suits were not located and the operator and crew are missing (unpublished Coast Guard situation reports, 1990).

The new EPIRB technology is impressive. Each Category 1 EPIRB will be electronically coded, thereby providing a means to immediately identify the vessel from which the signal is emanating. The signals will be received by satellites capable of fixing a signal position to within about 2 miles. A 121.5-MHz homing beacon that coactivates with the 406-MHz signal will permit localization by rescue aircraft once on scene. Test circuits are designed to indicate power status; however, there are no live transmissions of test signals. An automatic activating device will be required for Category 1 units. The technology is further designed to permit maintenance by authorized service centers only (see Embler, undated; Lemon, 1990b).

While impressive, the technology is not foolproof. It is possible to install some Category 1 EPIRBs in their float-free holders without arming them. Where a mercury-type switch is installed for automatic activation when the EPIRB is vertical, it remains possible to inadvertently activate the device if it is stored armed in a vertical position inside a vessel while moored. However, the position-fixing aspect of the new system and electronic, vessel-specific signature will make it easier to resolve such occurrences. Another potential problem for which limited information was available is the suitability of new Category 1 EPIRBs aboard smaller fishing vessels constructed of plastic with built-in flotation. Some EPIRBs may remain in their float-free holders if the vessel does not sink. Further examination of this issue was not feasible with available information.

Personal Locator Beacons (PLBs)

The personal equivalent of an EPIRB is a personal locator beacon (PLB). This device is most closely associated with transmitters worn in avalanche areas to alert rescuers to anyone buried in snow. PLBs are being considered by some countries for maritime use, but they are illegal in the United States except for a few restricted applications (Lemon, 1990a). The Federal Communications Commission recently warned a manufacturer to discontinue marketing an aircraft emergency locator transmitter (ELT) for PLB use.

INSTRUCTIONS FOR USING SURVIVAL EQUIPMENT

Regulatory Requirements for Instructions

Each manufacturer of a Coast Guard-approved PFD designed for recreational vessels is required to furnish an information pamphlet with each PFD sold (33 CFR 181). This pamphlet was recently upgraded to a nonalterable, 16-page booklet to overcome significant variations in format and content in previously published materials. A separate pamphlet is required for hybrid recreational PFDs (46 CFR 160). Because of the unique nature of recreational hybrids, formal donning and operating instructions must be included in the hybrid pamphlet. Other than hybrid PFDs, manufacturers of safety and survival equipment are required by Coast Guard regulation to provide written instructions only to purchasers of immersion suits. The Coast Guard has not required either informational pamphlets or written donning instructions for PFDs or other equipment sold for use aboard uninspected commercial vessels. This is partly because commercial users traditionally have been thought to have more maritime expertise than recreational boaters. However, where a PFD has been granted dual approval for recreational and commercial use, a copy of the recreational pamphlet must be provided.

Manufacturers are not required to provide instructional videotapes demonstrating proper care and use of PFDs, immersion suits, and EPIRBs, and they don't. Various generic manuals, handbooks, guides, and videotapes covering the care and proper use of survival equipment are available from private vendors, training institutions, and trade associations at reasonable to moderate costs (Hollin, 1982; Sabella, 1989). Several safety manuals with sections covering survival equipment are specifically tailored to the North Pacific and Gulf Coast (Sabella, 1986; Hollin and Middleton, 1989). An Atlantic Coast manual is being developed by the University of Rhode Island (Hollister and Carr, eds., 1990). A computer-aided audit program to track vessel maintenance and safety status is also available (Moran, 1989). Safety and survival courses are offered by academic and vocational institutions, trade associations, and private vendors (DeAlteris et al., 1989; Sabella, 1987; Pennington, 1987; Walker and Lodge, 1987; Keiffer, 1984).

Coast Guard Survival System Publications

The Coast Guard publishes a circular entitled *Voluntary Standards for U.S. Uninspected Commercial Fishing Vessels* (U.S. Coast Guard, 1986b). Commonly referred to as NVIC 5-86, this circular is a technical document covering fishing vessel design, construction, and equipment. It covers safety and survival equipment requirements, recommends equipment beyond minimum requirements, provides an overview of selected equipment performance, and offers guidelines on selected equipment maintenance. First published in August 1986, NVIC 5-86 was distributed to trade and fishermen associations. It remains available for purchase through the Coast Guard. No data were developed to indicate the extent to which NVIC 5-86 is available to operators and crews, and if available, how well the material is used in practice to improve the safety of fishing operations.

A comprehensive manual is published by the Coast Guard covering Coast Guard rescue and survival systems (U.S. Coast Guard, 1988a). It is widely available to Coast Guard forces. This manual provides comprehensive technical guidance on commercial and Navy survival equipment used by the Coast Guard as well as use guidance on operational use of selected equipment. Some of the commercial equipment used by the Coast Guard is identical to equipment found aboard some uninspected fishing vessels. The Coast Guard manual provides detailed maintenance guidance, while commercially available manuals tend to focus more on cursory checks that can be performed by persons with limited formal training.

APPENDIX

J

Commercial Fishing Industry Vessel Safety Act of 1988

PUBLIC LAW 100-424

PUBLIC LAW 100-424—SEPT. 9, 1988 102 STAT. 1585

Public Law 100-424
100th Congress

An Act

To provide for the establishment of additional safety requirements for fishing industry vessels, and for other purposes.

Sept. 9, 1988
[H.R. 1841]

Be it enacted by the Senate and House of Representatives of the United States of America in Congress assembled,

Commercial Fishing Industry Vessel Safety Act of 1988.
46 USC 2101 note.

SECTION 1. SHORT TITLE.

This Act may be cited as the "Commercial Fishing Industry Vessel Safety Act of 1988".

SEC. 2. UNINSPECTED COMMERCIAL FISHING INDUSTRY VESSEL SAFETY REQUIREMENTS.

(a) IN GENERAL.—Chapter 45 of title 46, United States Code, is amended to read as follows:

"CHAPTER 45—UNINSPECTED COMMERCIAL FISHING INDUSTRY VESSELS

"Sec.

"§ 4501. Application

46 USC 4501.

"(a) This chapter applies to an uninspected vessel which is a fishing vessel, fish processing vessel, or fish tender vessel.

"(b) This chapter does not apply to the carriage of bulk dangerous cargoes regulated under chapter 37 of this title.

"§ 4502. Safety standards

46 USC 4502.
Regulations.

"(a) The Secretary shall prescribe regulations which require that each vessel to which this chapter applies shall be equipped with—

"(1) readily accessible fire extinguishers capable of promptly and effectively extinguishing a flammable or combustible liquid fuel fire;

"(2) at least one readily accessible life preserver or other lifesaving device for each individual on board;

"(3) an efficient flame arrestor, backfire trap, or other similar device on the carburetors of each inboard engine which uses gasoline as fuel;

"(4) the means to properly and efficiently ventilate enclosed spaces, including engine and fuel tank compartments, so as to remove explosive or flammable gases;

"(5) visual distress signals;

"(6) a buoyant apparatus, if the vessel is of a type required by regulations prescribed by the Secretary to be equipped with that apparatus;

"(7) alerting and locating equipment, including emergency position indicating radio beacons, on vessels that operate on the high seas; and

"(8) a placard as required by regulations prescribed under section 10603(b) of this title.

Regulations.

"(b) In addition to the requirements of subsection (a) of this section, the Secretary shall prescribe regulations for documented vessels to which this chapter applies that operate beyond the Boundary Line or that operate with more than 16 individuals on board, for the installation, maintenance, and use of—

"(1) alerting and locating equipment, including emergency position indicating radio beacons;

"(2) lifeboats or liferafts sufficient to accommodate all individuals on board;

"(3) at least one readily accessible immersion suit for each individual on board that vessel when operating on the waters described in section 3102 of this title;

"(4) radio communications equipment sufficient to effectively communicate with land-based search and rescue facilities;

"(5) navigation equipment, including compasses, radar reflectors, nautical charts, and anchors;

"(6) first aid equipment, including medicine chests; and

"(7) other equipment required to minimize the risk of injury to the crew during vessel operations, if the Secretary determines that a risk of serious injury exists that can be eliminated or mitigated by that equipment.

"(c) In addition to the requirements described in subsections (a) and (b) of this section, the Secretary may prescribe regulations establishing minimum safety standards for vessels to which this chapter applies that were built after December 31, 1988, or that undergo a major conversion completed after that date, and that operate with more than 16 individuals on board, including standards relating to—

"(1) navigation equipment, including radars and fathometers;

"(2) life saving equipment, immersion suits, signaling devices, bilge pumps, bilge alarms, life rails, and grab rails;

"(3) fire protection and firefighting equipment, including fire alarms and portable and semiportable fire extinguishing equipment;

"(4) use and installation of insulation material;

"(5) storage methods for flammable or combustible material; and

"(6) fuel, ventilation, and electrical systems.

Regulations.

"(d)(1) The Secretary shall prescribe regulations for the operating stability of a vessel to which this chapter applies—

"(A) that was built after December 31, 1989; or

"(B) the physical characteristics of which are substantially altered after December 31, 1989, in a manner that affects the vessel's operating stability.

"(2) The Secretary may accept, as evidence of compliance with this subsection, a certification of compliance issued by the person providing insurance for the vessel or by another qualified person approved by the Secretary.

"(e) In prescribing regulations under this chapter, the Secretary—

"(1) shall consider the specialized nature and economics of the operations and the character, design, and construction of the vessel; and

"(2) may not require the alteration of a vessel or associated equipment that was constructed or manufactured before the effective date of the regulation.

"(f) The Secretary shall examine a fish processing vessel at least once every two years to ensure that the vessel complies with the requirements of this chapter.

"§ 4503. Fish processing vessel certification

46 USC 4503.

"(a) A fish processing vessel to which this section applies may not be operated unless the vessel—

"(1) meets all survey and classification requirements prescribed by the American Bureau of Shipping or another similarly qualified organization approved by the Secretary; and

"(2) has on board a certificate issued by the American Bureau of Shipping or that other organization evidencing compliance with this subsection.

"(b) This section applies to a fish processing vessel to which this chapter applies that—

"(1) is built after July 27, 1990; or

"(2) undergoes a major conversion completed after that date.

"§ 4504. Prohibited acts

46 USC 4504.

"A person may not operate a vessel in violation of this chapter or a regulation prescribed under this chapter.

"§ 4505. Termination of unsafe operations

46 USC 4505.

"An official authorized to enforce this chapter—

"(1) may direct the individual in charge of a vessel to which this chapter applies to immediately take reasonable steps necessary for the safety of individuals on board the vessel if the official observes the vessel being operated in an unsafe condition that the official believes creates an especially hazardous condition, including ordering the individual in charge to return the vessel to a mooring and to remain there until the situation creating the hazard is corrected or ended; and

"(2) may order the individual in charge of an uninspected fish processing vessel that does not have on board the certificate required under section 4503(1) of this title to return the vessel to a mooring and to remain there until the vessel is in compliance with that section.

"§ 4506. Exemptions

46 USC 4506.

"(a) The Secretary may exempt a vessel from any part of this chapter if, under regulations prescribed by the Secretary (including regulations on special operating conditions), the Secretary finds that—

"(1) good cause exists for granting an exemption; and

"(2) the safety of the vessel and those on board will not be adversely affected.

"(b) A vessel to which this chapter applies is exempt from section 4502(b)(2) of this title if it—

"(1) is less than 36 feet in length; and

"(2) is not operating on the high seas.

46 USC 4507

"§ 4507. Penalties

"(a) The owner, charterer, managing operator, agent, master, and individual in charge of a vessel to which this chapter applies which is operated in violation of this chapter or a regulation prescribed under this chapter may each be assessed a civil penalty by the Secretary of not more than $5,000. Any vessel with respect to which a penalty is assessed under this subsection is liable in rem for the penalty.

"(b) A person willfully violating this chapter or a regulation prescribed under this chapter shall be fined not more than $5,000, imprisoned for not more than one year, or both.

46 USC 4508. Establishment.

"§ 4508. Commercial Fishing Industry Vessel Advisory Committee

"(a) The Secretary shall establish a Commercial Fishing Industry Vessel Advisory Committee. The Committee—

"(1) may advise, consult with, report to, and make recommendations to the Secretary on matters relating to the safe operation of vessels to which this chapter applies, including navigation safety, safety equipment and procedures, marine insurance, vessel design, construction, maintenance and operation, and personnel qualifications and training;

"(2) may review proposed regulations under this chapter;

"(3) may make available to Congress any information, advice, and recommendations that the Committee is authorized to give to the Secretary; and

"(4) shall meet at the call of the Secretary, who shall call such a meeting at least once during each calendar year.

"(b)(1) The Committee shall consist of seventeen members with particular expertise, knowledge, and experience regarding the commercial fishing industry as follows:

"(A) ten members from the commercial fishing industry who—

"(i) reflect a regional and representational balance; and

"(ii) have experience in the operation of vessels to which this chapter applies or as a crew member or processing line worker on an uninspected fish processing vessel;

"(B) three members from the general public, including, whenever possible, an independent expert or consultant in maritime safety and a member of a national organization composed of persons representing owners of vessels to which this chapter applies and persons representing the marine insurance industry;

"(C) one member representing each of—

"(i) naval architects or marine surveyors;

"(ii) manufacturers of equipment for vessels to which this chapter applies;

"(iii) education or training professionals related to fishing vessel, fish processing vessel, or fish tender vessel safety or personnel qualifications; and

"(iv) underwriters that insure vessels to which this chapter applies.

Federal Register, publication.

"(2) At least once each year, the Secretary shall publish a notice in the Federal Register and in newspapers of general circulation in coastal areas soliciting nominations for membership on the Committee, and, after timely notice is published, appoint the members of the Committee. An individual may be appointed to a term as a member of the Committee more than once.

"(3)(A) A member of the Committee shall serve a term of three years.

"(B) If a vacancy occurs in the membership of the Committee, the Secretary shall appoint a member to fill the remainder of the vacated term.

"(4) The Committee shall elect one of its members as the Chairman and one of its members as the Vice Chairman. The Vice Chairman shall act as Chairman in the absence or incapacity of, or in the event of a vacancy in the office of, the Chairman.

"(5) The Secretary shall, and any other interested agency may, designate a representative to participate as an observer with the Committee. These representatives shall, as appropriate, report to and advise the Committee on matters relating to vessels to which this chapter applies which are under the jurisdiction of their respective agencies. The Secretary's designated representative shall act as executive secretary for the Committee and perform the duties set forth in section 10(c) of the Federal Advisory Committee Act (5 App. U.S.C.).

"(c)(1) The Secretary shall, whenever practicable, consult with the Committee before taking any significant action relating to the safe operation of vessels to which this chapter applies.

"(2) The Secretary shall consider the information, advice, and recommendations of the Committee in consulting with other agencies and the public or in formulating policy regarding the safe operation of vessels to which this chapter applies.

"(d)(1) A member of the Committee who is not an officer or employee of the United States or a member of the Armed Forces, when attending meetings of the Committee or when otherwise engaged in the business of the Committee, is entitled to receive—

"(A) compensation at a rate fixed by the Secretary, not exceeding the daily equivalent of the current rate of basic pay in effect for GS-18 of the General Schedule under section 5332 of title 5 including travel time; and

"(B) travel or transportation expenses under section 5703 of title 5.

"(2) Payments under this section do not render a member of the Committee an officer or employee of the United States or a member of the Armed Forces for any purpose.

"(3) A member of the Committee who is an officer or employee of the United States or a member of the Armed Forces may not receive additional pay based on the member's service to the Committee.

"(4) The provisions of this section relating to an officer or employee of the United States or a member of the Armed Forces do not apply to a member of a reserve component of the Armed Forces unless that member is in an active status.

Termination date.

"(e)(1) The Federal Advisory Committee Act (5 U.S.C. App. 1 et seq.) applies to the Committee, except that the Committee terminates on September 30, 1992.

"(2) Two years prior to the termination date referred to in paragraph (1) of this subsection, the Committee shall submit to Congress its recommendation regarding whether the Committee should be renewed and continued beyond the termination date.".

46 USC 4508 note.

(b) INITIAL APPOINTMENTS TO COMMERCIAL FISHING INDUSTRY ADVISORY COMMITTEE.—

(1) TERMS OF INITIAL APPOINTMENTS.—Of the members first appointed to the Commercial Fishing Industry Advisory Committee under section 4508 of title 46, United States Code (as amended by this Act)—

(A) one-third of the members shall serve a term of one year and one-third of the members shall serve a term of two years, to be determined by lot at the first meeting of the Committee; and

(B) terms may be adjusted to coincide with the Government's fiscal year.

(2) COMPLETION OF INITIAL APPOINTMENTS.—The Secretary shall complete appointment of members pursuant to this subsection not later than 90 days after the date of the enactment of this Act.

(c) REPEAL.—Subsection (e) of section 4102 of title 46, United States Code, is repealed.

46 USC 7101 note.

SEC. 3. PLAN FOR LICENSING OPERATORS OF FISHING INDUSTRY VESSELS.

The Secretary of the department in which the Coast Guard is operating shall, within two years after the date of enactment of this Act, and in close consultation with the Commercial Fishing Industry Vessel Advisory Committee established under section 4508 of title 46, United States Code (as amended by this Act), prepare and submit to the Congress a plan for the licensing of operators of documented fishing, fish processing, and fish tender vessels. The plan shall take into consideration the nature and variety of the different United States fisheries and of the vessels engaged in those fisheries, the need to license all operators or only those working in certain types of fisheries or vessels, and other relevant factors.

SEC. 4. ACCIDENT DATA STATISTICS.

(a) COMPILATION AND SUBMISSION OF DATA.—Chapter 61 of title 46, United States Code, is amended by adding at the end the following new section:

Records.
Insurance.
46 USC 6104.

"§ 6104. Commercial fishing industry vessel casualty statistics

"(a) The Secretary shall compile statistics concerning marine casualties from data compiled from insurers of fishing vessels, fish processing vessels, and fish tender vessels.

"(b) A person underwriting primary insurance for a fishing vessel, fish processing vessel, or fish tender vessel shall submit periodically to the Secretary data concerning marine casualties that is required by regulations prescribed by the Secretary.

Regulations.

"(c) After consulting with the insurance industry, the Secretary shall prescribe regulations under this section to gather a statistical base for analyzing vessel risks.

"(d) The Secretary may delegate to a qualified person that has knowledge and experience in the collection of statistical insurance data the authority of the Secretary under this section to compile statistics from insurers.".

(b) PENALTY.—Section 6103 of title 46, United States Code, is amended as follows:

(1) before "An" insert "(a)"; and

(2) add the following new subsection:

"(b) A person failing to comply with section 6104 of this title or a regulation prescribed under that section is liable to the Government for a civil penalty of not more than $5,000.".

(c) CONFORMING AMENDMENT.—The analysis for chapter 61 of title 46, United States Code, is amended by adding at the end the following:

"6104. Commercial fishing industry vessel casualty statistics.".

SEC. 5. STUDIES.

46 USC 4502 note.

(a) FISHING INDUSTRY VESSEL INSPECTION STUDY.—The Secretary of Transportation, utilizing the National Academy of Engineering and in consultation with the National Transportation Safety Board, the Commercial Fishing Industry Vessel Advisory Committee, and the fishing industry, shall—

(1) conduct a study of the safety problems on fishing industry vessels;

(2) make recommendations regarding whether a vessel inspection program should be implemented for fishing vessels, fish tender vessels, and fish processing vessels, including recommendations on the nature and scope of that inspection; and

(3) submit the study and recommendations to Congress before January 1, 1990.

(b) UNCLASSIFIED FISH PROCESSING VESSEL STUDY.—The Secretary of the department in which the Coast Guard is operating, in consultation with the Commercial Fishing Industry Vessel Advisory Committee established under section 4508 of title 46, United States Code (as amended by this Act), and with representatives of persons operating fish processing vessels—

(1) shall conduct a study of fish processing vessels that are not surveyed and classed by an organization approved by the Secretary;

(2) shall make recommendations regarding what hull and machinery requirements should apply to vessels described in paragraph (1) to ensure that those vessels are operated and maintained in a condition in which they are safe to operate at sea; and

(3) shall submit the study and recommendations to Congress before July 28, 1991.

SEC. 6. FISHING VOYAGE REQUIREMENTS.

(a) ENACTMENT OF NEW CHAPTER IN TITLE 46.—Title 46, United States Code, is amended by inserting after chapter 105 the following:

"CHAPTER 106—FISHING VOYAGES

"§ 10601. Fishing agreements

46 USC 10601.

"(a) Before proceeding on a voyage, the master or individual in charge of a fishing vessel, fish processing vessel, or fish tender vessel shall make a fishing agreement in writing with each seaman enployed on board if the vessel is—

"(1) at least 20 gross tons; and

"(2) on a voyage from a port in the United States.

"(b) The agreement shall be signed also by the owner of the vessel.

"(c) The agreement shall—

"(1) state the period of effectiveness of the agreement;

Wages.

"(2) include the terms of any wage, share, or other compensation arrangement peculiar to the fishery in which the vessel will be engaged during the period of the agreement; and

"(3) include other agreed terms.

46 USC 10602.

"§ 10602. Recovery of wages and shares of fish under agreement

"(a) When fish caught under an agreement under section 10601 of this title are delivered to the owner of the vessel for processing and are sold, the vessel is liable in rem for the wages and shares of the proceeds of the seamen. An action under this section must be brought within six months after the sale of the fish.

"(b)(1) In an action under this section, the owner shall produce an accounting of the sale and division of proceeds under the agreement. If the owner fails to produce the accounting, the vessel is liable for the highest value alleged for the shares.

"(2) The owner may offset the value of general supplies provided for the voyage and other supplies provided the seaman bringing the action.

"(c) This section does not affect a common law right of a seaman to bring an action to recover the seaman's share of the fish or proceeds.

46 USC 10603.

"§ 10603. Seaman's duty to notify employer regarding illness, disability, and injury

"(a) A seaman on a fishing vessel, fish processing vessel, or fish tender vessel shall notify the master or individual in charge of the vessel or other agent of the employer regarding any illness, disability, or injury suffered by the seaman when in service to the vessel not later than seven days after the date on which the illness, disability, or injury arose.

Regulations.

"(b) The Secretary shall prescribe regulations requiring that each fishing vessel, fish processing vessel, and fish tender vessel shall have on board a placard displayed in a prominent location accessible to the crew describing the seaman's duty under subsection (a) of this section.".

(b) CONFORMING AMENDMENT.—The table of contents at the beginning of title 46, United States Code, is amended by inserting after the item relating to chapter 105 the following:

"106. Fishing voyages..10601.".

(c) REPEALS.—Sections 4391, 4392, 4393, and 4394 of the Revised Statutes of the United States (46 App. U.S.C. 531-534) are repealed.

46 USC 4501 note. Termination date.

SEC. 7. TRANSITIONAL PROVISION.

Until July 28, 1990, a foreign built fish processing vessel subject to chapter 45 of title 46, United States Code, is deemed to comply with the requirements of that chapter if—

(1) it has an unexpired certificate of inspection issued by a foreign country that is a party to an International Convention for Safety of Life at Sea to which the United States Government is a party; and

(2) it is in compliance with the safety requirements of that foreign country that apply to that vessel.

SEC. 8. TECHNICAL AND CONFORMING AMENDMENTS.

(a) IMMERSION SUITS.—

(1) REQUIREMENT.—Section 3102 of title 46, United States Code, is amended by striking "exposure" each place it appears and inserting in lieu thereof "immersion".

(2) SECTION HEADING.—The section heading for section 3102 of that title is amended by striking "**Exposure**" and inserting in lieu thereof "**Immersion**".

(3) ANALYSIS.—The chapter analysis for chapter 31 of that title is amended by striking "Exposure" and inserting in lieu thereof "Immersion".

(b) OTHER UNINSPECTED VESSEL REQUIREMENTS.—Section 4101 of title 46, United States Code, is amended by inserting "not subject to chapter 45 of this title" after "uninspected vessel".

(c) MAJOR CONVERSION DEFINED.—

(1) DEFINITION.—Section 2101 of title 46, United States Code, is amended by inserting after paragraph (14) the following:

"(14a) 'major conversion' means a conversion of a vessel that—

"(A) substantially changes the dimensions or carrying capacity of the vessel;

"(B) changes the type of the vessel;

"(C) substantially prolongs the life of the vessel; or

"(D) otherwise so changes the vessel that it is essentially a new vessel, as decided by the Secretary.".

(2) REPEAL.—Section 3701(2) of title 46, United States Code, is repealed.

Approved September 9, 1988.

LEGISLATIVE HISTORY—H.R. 1841:

HOUSE REPORTS: No. 100-729 (Comm. on Merchant Marine and Fisheries).
CONGRESSIONAL RECORD, Vol. 134 (1988):
June 27, considered and passed House.
Aug. 11, considered and passed Senate.

Index

D

E

G

H

I

J

L

O

P

T

W